Applied Electron Microscopy
Angewandte Elektronenmikroskopie

Band 9

Applied Electron Microscopy
Angewandte Elektronenmikroskopie

Band 9

Prof. Dr. Josef Zweck (Hrsg.)

Taryl L. Kirk

Near Field Emission Scanning Electron Microscopy

Logos Verlag Berlin

λογος

Applied Electron Microscopy
Angewandte Elektronenmikroskopie

Herausgeber:
Prof. Dr. Josef Zweck
Institut für Experimentelle und Angewandte Physik
Universität Regensburg
93040 Regensburg
Germany
Email: josef.zweck@physik.uni-regensburg.de

Bibliografische Information der Deutschen Nationalbibliothek

Die Deutsche Nationalbibliothek verzeichnet diese Publikation in der Deutschen Nationalbibliografie; detaillierte bibliografische Daten sind im Internet über http://dnb.d-nb.de abrufbar.

ISBN 978-3-8325-2518-7
ISSN 1860-0034

Logos Verlag Berlin GmbH
Comeniushof, Gubener Str. 47,
10243 Berlin

Tel.: +49 (0)30 / 42 85 10 90
Fax: +49 (0)30 / 42 85 10 92
http://www.logos-verlag.de

Diss. ETH Nr. 18824

Near Field Emission Scanning Electron Microscopy

A dissertation submitted to the
ETH Zurich

for the degree of
Doctor of Sciences

presented by

Taryl Leaton Kirk
M.Sc. in Physics, Universität Stuttgart
born on July 14, 1978
citizen of the Republic of Trinidad and Tobago and the United States of America

accepted on the recommendation of

Prof. Dr. Danilo Pescia, examiner
Dr. Richard G. Forbes, co-examiner

2010

dedicated to my sons, Vincent and Victor, and my wife Meike

"Since my journey into tunnelling is still continuing . . . However, I would like to point out that many high barriers exist in this world: Barriers between nations, races and creeds. Unfortunately, some barriers are thick and strong. But I hope, with determination, we will find a way to tunnel through these barriers easily and freely, to bring the world together . . ."

Nobel Lecture, December 12, 1973

By *Leo Esaki*

私のトンネル効果への旅はまだ続いていますから，この話がつきることはありませんが，この狭くなってきた私たちの住む地球上にはなお多くの高い障壁が現存することを指摘したいと思います。それは国家間，人種間，そしてまた信条教義を異にする人達のあいだにある障壁です。不幸にしてある障壁は厚く強固であります。私たちの熱意によって，これらの障壁を自由に透過することができる新しいトンネル効果をみいだすことができれば，それこそ世界の永遠平和の礎となり，アルフレッド・ノーベルの遺言に添うことができると思います。

Contents

Nahfeldemissions-Rasterelektronenmikroskopie

In dieser Dissertation zeige ich die Entwicklung und die Anwendung eines neuartigen Mikroskops, des Nahfeldemissions-Rasterelektronenmikroskops (NFEREM) auf. Die Bezeichnung Nahfeld deutet auf die Tatsache hin, dass die Elektronenquelle bei einem Abstand von nur einigen 10 nm von der abzubildenden Oberfläche entfernt liegt. Dieses Merkmal unterscheidet dieses Mikroskop von den konventionellen Rasterelektronenmikroskopen, bei denen die Quelle einige 10 cm von der abzubildenen Oberfläche entfernt ist. Darüber hinaus sei bemerkt, dass dies kein optisches Mikroskop ist, in dem das Bild der Oberfläche durch Lichteinstrahlung, Lichtreflektion oder Lichtstreuung erzeugt wird.

Nach der Einführung (Kapitel 1) betrachtet Kapitel 2 die theoretischen Grundlagen der Rastertunnelmikroskopie (*RTM*). In der *RTM* wird eine metallische Nadel so nahe (weniger als 1 nm) an die Oberfläche gebracht, dass tunneln zwischen der Nadel und dem naheliegendsten Oberflächenatom möglich wird. Sofern die Nadel auf Abstände von einigen 10 nm von der Oberfläche zurückgezogen wird, können einige Elektronen durch das Phänomen der Feldemission die Nadel verlassen. Ferner können diese Elektronen mit wählbarer Energie auf die naheliegende Oberfläche beschleunigt werden, wo sie durch inelastische und elastische Elektronenstreuungen weitere Elektronen zum Verlassen der Oberfläche anregen können. Des Weiteren diskutiere ich im Kapitel 2 die theoretischen Grundlagen des Feldemissionprozess und der Elektronenstreuung.

Kapitel 3 erläutert die technischen Details der Proben- und Nadelvorbereitung sowie den technischen Aufbau des ad hoc entwickelten Instruments. Der wesentliche technische Aspekt dieses Instruments ist die Tatsache [siehe R. Palmer et al. Appl. Phys. Lett. **77**, 4223 (2000)], dass die angeregten Elektronen vorzüglich parallel zur Oberfläche austreten.

In Kapitel 4 stelle ich die Resultate der Experimente vor. Durch das Rastern der Nadel bei **konstanter Entfernung** von der Oberfläche wird sowohl der Feldemissionsstrom als auch die Intensität der aus der Oberfläche emittierten Elektronen durch die Topographie der Oberfläche moduliert. Infolgedessen entstehen zwei gleichzeitige Bilder der Oberfläche. Erwähnenswerterweise basiert eines der Bilder auf feldemittierten Elektronen und eines der Bilder auf sekundären Elektronen. Bei meinen Experimenten wurde eine vertikale räumliche Auflösung von etwa 0.1 nm und eine laterale räumliche Auflösung von einigen nm erreicht und monoatomare Stufen auf W (110) sichtbar gemacht.

Im Anhang A.1, A.2, und A.3 präsentiere ich eine Auswahl von rohen Messdaten, um die Grenzen von *NFEREM* zu illustrieren und um den Vergleich mit *RTM* zu ermöglichen.

Near field emission scanning electron microscopy

This dissertation presents the development and application of a novel microscope, the "Near Field Emission Scanning Electron Microscope" (*NFESEM*). Here, the term "near" references the fact that the electron source is located at a few tens of nanometers from the sample surface that is under investigation. This characteristic distinguishes the present instrument from more conventional *SEM*s, which use "remote" field-emitted electron sources. Moreover, this is not an optical measurement, such as in scanning near-field optical microscopy, in which an image is generated by exciting and collecting **light** scattered in the "near-field" regime.

After a brief introduction (Chapter 1), Chapter 2 treats the fundamental theoretical aspects of Scanning Tunneling Microscopy (*STM*). In *STM* on metal surfaces, a metallic tip is brought so close to a surface (less than 1nm) that tunneling can occur between the tip and the surface atom residing closest to the tip. Therefore, *STM* is one of the methods of choice for achieving atomic spatial resolution. When the tip is retracted to distances of some tens of nanometers, electrons are ejected from the tip via the process of field emission (*FE*). These electrons are accelerated with variable energy toward the underlying surface, where, by elastic and inelastic scattering processes, they can excite electrons to leave the surface. Subsequently, Chapter 2 treats the fundamental aspects of the *FE* process in the near *FE* regime and of the secondary electron (*SE*) production.

Chapter 3 provides details about the sample and the tip preparation, in addition to the technical aspects of the home-built instrument. The key technical result is that, in agreement with the results by R. Palmer *et al.* [Appl. Phys. Lett. **77**, 4223 (2000)], *SE*s, excited in the near field configuration, are preferentially ejected parallel to the surface.

Chapter 4 presents the experimental results. Upon scanning the tip over the surface in **constant height** (***CH***) mode, both the *FE* current and the intensity of the electrons emitted from the surface are modulated by the surface topography. As a consequence, images of the surface topography can be obtained. A particular feature of the present *NFESEM* is that two separate but simultaneously-generated images are produced: one based on the *FE* current and the other based on the variations in the *SE* detector signal. In regards to imaging, this *NFESEM* study has achieved a vertical resolution of about 0.2 nm and an ultimate lateral resolution of $<$ 2nm, which is able to resolve the step edges on a W (110) substrate.

In the Appendices A.1, A.2, and A.3 , "raw" experimental data are presented, with the aim of acquainting the reader with the limitations of *NFESEM*, and also compares its performance to *STM*.

1. Introduction

E. Ruska was awarded the Noble Prize in 1986 for his work toward the development of microscopes that were based on a primary beam consisting of electrons instead of light. In the same year, H. Rohrer and G. Binnig were also awarded the Nobel Prize for their discovery of an alternative method to perform ultra-high resolution microscopy: *STM* (scanning tunneling microscopy). This technique is based on the quantum mechanical tunneling effect, which limits the interaction of a tip (brought to sub-nanometer distances from a surface) to the surface atom residing nearest the tip, thus achieving the elusive atomic resolution.

The scope of this dissertation is to combine the fundamental concepts of these two types of microscopy, thereby creating a quite compact instrument that can be integrated easily with pre-existing systems. The close proximity between the source and the object provides a means of overcoming the limitations of conventional *SEM* and opens the possibility to use lower primary beam energies. Although these experiments conceptually resemble Binnig and Rohrer's *STM*, they generate images similar to *SEM* without the need of a remote electron gun.

This instrument (Chapter 3, devoted to experimental details) uses low energies (typically a few tens of electron volts), and the electron source, remote in standard *SEM*s, is brought within tens of nanometers to the object. This method is called "near field emission scanning electron microscopy" (*NFESEM*). The terminology "near" refers to the locality of the electron source and is used to distinguish this instrument from those – more standard *SEM*s – using "remote" field emission gun electron sources. It is important to note that this is not an optical measurement, such as in scanning near-field optical microscopy.

NFESEM uses electron sources based on cold field emission (*CFE*) (Chapter 2, a discussion of the theoretical foundations). *CFE* occurs when a metallic tip is submerged in strong electric fields that alter the potential barrier, at the tip surface, permitting electrons near the Fermi level to tunnel through it, in accordance with the Fowler-Nordheim *CFE* equation. These electrons act as the primary beam, which in conventional *SEM* originates from a remote source. They impinge on a well-defined area on the surface, localized by the tip, and excite secondary electrons (*SE*)s that are subsequently collected and counted by a detector. R. D. Young *et al.* were in fact the first to employ a *CFE* current between a sharp metallic tip and a conducting sample to resolve surface features, as small as 400 nm. H.-W. Fink later achieved a lateral resolution of 3 nm on a polycrystalline gold surface using a similar technique with a single atom (111)-oriented tungsten (*W*)-tip.

Contrary to these earlier works, the present method operates in constant height (*CH*) mode, *i.e.* the tip is scanned at a fixed height while the field-emitted current is allowed to vary with the sample surface topography. Accordingly, the *FE* current also depends on the surface topography, and this allows for two simultaneous images of the surface to be acquired: one based on the modulation of the *FE* current, and the other based on the intensity of the *SE*s emitted from the surface (see Chapter 4, discussing the main results). The purely *SE* characteristics, *e.g.* surface magnetization and/or chemical contrasts might be extracted by comparing *FE* and *SE* images. A follow-up instrument, which allows for the spin polarization of the *SE*s to be measured, is now in the process of development. A remarkable result of this dissertation (see Chapter 4 and the "raw" experimental data shown in the Appendix A.3) is the spatial resolution achieved by the instrument: vertical to the surface, step edges have been resolved, indicating a vertical resolution of about 0.2 nm. Laterally, a spatial resolution of a few nanometers has been achieved, which, taking into account that the tip-surface distance is few tens of nanometers, is a surprising result. One of the main future tasks is the understanding of the spatial resolution capabilities of this instrument.

2. Theory

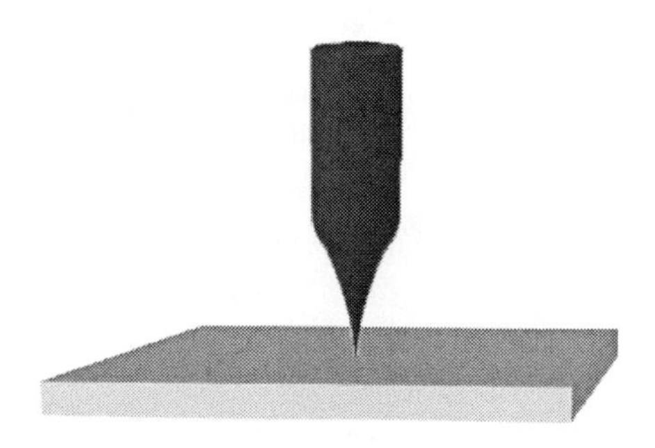

Figure 2.1.: Schematic of a point-contact on a metallic surface.

Near field emission scanning electron microscopy (*NFESEM*) is a relatively recent discovery, and requires prior knowledge of similar measurement techniques in order to fully grasp the functionality of the microscope. In this chapter the theory of *NFESEM* will be motivated by a discussion of the theory of a related microscopic technique, namely scanning tunneling microscopy (*STM*), with the aim of stressing the fundamental physical differences between the *NFESEM* and *STM*. In *STM*, the spatial separation between tip and surface is in the subnanometer range, allowing for electrons to tunnel through vacuum between the orbital states of the probe and the sample. In *NFESEM*, electrons tunnel from the tip into the vacuum and are subsequently accelerated towards the sample, striking it and causing electrons to be ejected from the surface.

Alternatively, another type of local surface probe measurement, known as point-contact spectroscopy (*PCS*), can be performed. This kind of surface probe involves the tip being in physical contact with the surface. Electronic transport occurs via a small metallic constriction between two, typically metallic, conductors (Fig. 2.1) established by the impingement of a needle-shaped electrode on a planar surface. Sharvin [1] was the first to investigate the properties of metals with a large mean free path of electrons, and found that the electron transport essentially follows Ohm's law. Some small non-linearities, observed in the second derivative of the current with respect to the applied voltage, are used to extract the local density of states of excitations, *e.g.* phonons and magnons, responsible for electron scattering at the location of the constriction *PCS*.[1] Although *PCS* is a powerful tool for the study of ballistic electrons flowing through a confined region on the specimen, it offers no method to image the surface under investigation, from a topographic point of view.

[1] In our group, T. Michlmayr has successfully implemented this technique to produce a local magnetic field in the vicinity of the point contact [2], thus advancing our knowledge of tip functionality.

Topographic imaging of the surface occurs when the tip is slightly retracted to a position where it is no longer in mechanical contact with the sample, in the classical sense. Only the wave functions of the orbitals overlap in this regime, allowing for electrons to flow in accordance with the density of states of the tip and the sample (see Fig. 2.2). As this is a quantum mechanical phenomenon, it will be discussed from a quite basic perspective of quantum tunneling. Further retraction of the tip brings about the regime of *NFESEM*. It is important to note that, in *NFESEM*, the impinging electrons define an interaction volume within the top surface layers. Inside this volume, secondary electrons are produced through mainly electron-electron scattering. These processes must be taken into account when defining the spatial resolution of the *NFESEM* and to determine the origin of the contrast mechanism, and will be briefly discussed in this chapter. Special attention will be paid to the most plausible scattering models and its implementation. Monte Carlo simulations for the associated materials reveal that the interaction volume is slightly larger than the beam diameter with a penetration depth less than 1 nm.

2.1. Fundamental aspects of STM

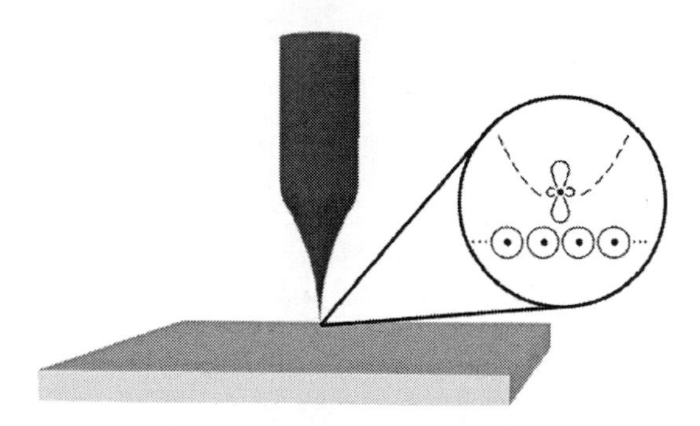

Figure 2.2.: Schematic of a needle electrode in tunneling contact with a metallic surface.

Quantum mechanics provides us with a world, in which phenomena that are known to be forbidden can actually be realized. One fascinating example of these phenomena is the tunneling of particles, classically forbidden, through potential barriers. This is in part due to the dual nature of particles, having both properties of rigid bodies and also of waves. Quantum mechanics can describe the motion of particles by defining them as plane waves. The plane wave description of a particle in one dimension can be expressed as:

$$\Psi\left(x\right) = Ae^{\pm ikx}; \tag{2.1}$$

where A is the amplitude of the wave and k is the wave number, which is associated with the energy E of the particle($k^2 = 2m_eE/\hbar$). The elastic tunneling effect can be analyzed by applying a one dimensional, time-independent Schrödinger equation to this wave function:

$$\left[\frac{1}{2m_e}\left(\frac{\hbar}{i}\frac{\partial}{\partial x}\right)^2 + V\left(x\right)\right]\Psi\left(x\right) = E\Psi\left(x\right). \tag{2.2}$$

A non-zero rectangular potential barrier can be defined in a specified region, which is thin enough for the particle to tunnel through with a high probability of being found on the other side of the barrier.

$$V(x) = \begin{cases} 0 \; for \; x \notin [0, L] & (outside \; of \; the \; barrier) \\ V_0 \; for \; x \in [0, L] & (within \; the \; barrier) \end{cases} \tag{2.3}$$

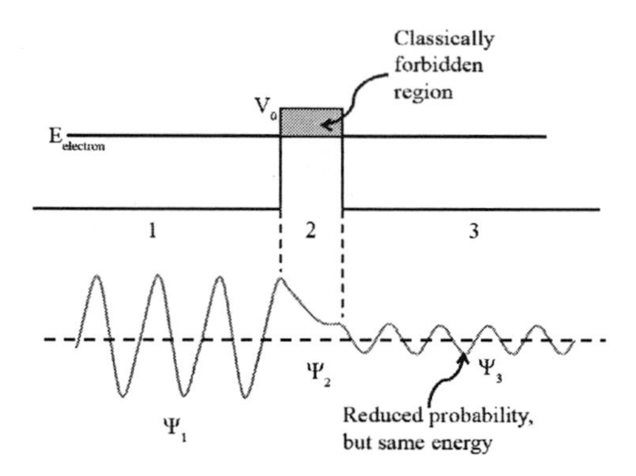

Figure 2.3.: Schematic of quantum tunneling through a rectangular non-zero potential barrier.

In Fig. 2.3), are the solutions for the real part of the three different regions, (1) before the barrier, (2) within the barrier, and (3) after the barrier.

To the left of the barrier, in region (1), it is convenient to consider plane wave functions that are localized within a rectangular three-dimensional box of volume Ω. Thus, the incident wave function has the "normalized" form $\Psi_i = \Omega^{-1/2}e^{ikx}$. So the full solution in region (1) can be written:

$$\Psi_1 = \Omega^{-1/2}\left(e^{-ikx} + C_1 e^{ikx}\right), \tag{2.4a}$$

Within the barrier in region (2):

$$\Psi_2 = \Omega^{-1/2}\left(B_2 e^{-\kappa x} + C_2 e^{\kappa x}\right) \; where \; \kappa^2 = \frac{2m_e\left(V_0 - E\right)}{\hbar^2}, \tag{2.4b}$$

To the right of the barrier in region (3), it is assumed that only an outgoing solution exists

$$\Psi_3 = \Omega^{-1/2} B_3 e^{ikx}. \tag{2.4c}$$

The solutions indicate that, at each boundary, the wave is partially reflected and transmitted.

The transmission coefficient, D, is the ratio of the transmitted probability current density to the incident probability current density, and is found by properly matching the wave functions at the boundaries. In general, the probability current density can be expressed as:

$$D \doteq \frac{|C_3|^2}{|C_1|^2} = \frac{1}{1 + (k^2+\kappa^2)^2 / (4k^2\kappa^2)\, sinh^2(\kappa L)} \quad (2.5)$$

$$\approx \frac{16k^2\kappa^2}{(k^2+\kappa^2)^2} \cdot e^{-2\kappa L}, \quad (2.6)$$

(the approximate value being for $\kappa L \gg 1$). This shows that the transmission coefficient has a strong exponential dependence on the barrier length and height. For example, for a barrier of height equal to a few electron-volts, increasing the thickness from 5 Å to 6 Å, decreases the exponential factor (and hence D) by an order of magnitude. The transmission coefficient can be computed for a general potential well using the Jeffreys, Wentzel, Kramers, and Brillouin ($JWKB$) semi-classical approximation:

$$D \approx \exp\left[-2\sqrt{2m_e}/\hbar \int \sqrt{V(x)-E}dx\right]. \quad (2.7)$$

The sensitivity of the transmission coefficient led experimentalist to devise an instrument that can use this to their advantage, which marked the advent of scanning tunneling microscopy (*STM*). This technique is explained in the following energy scheme. The sample is on the left hand side and is generally metallic with a given Fermi energy E_F. The tip is on the right hand side and its energy scheme is shifted down with a bias voltage V_B by an amount eV_B. In virtue of the finite quantum mechanical transmission coefficient, electrons from the occupied bands in the sample can tunnel through vacuum into the unoccupied bands on the tip side, leading to a net current flow:

$$I(eV_B) \propto \int_{E_F}^{E_F+eV_B} n_{tip}(E)\, n_{Sample}(E) \cdot D(E)\, dE. \quad (2.8)$$

Essentially, the tunnel current probes the transmission coefficient and the convolution of the **local** electronic density of states (*LDOS*) in the tip and the sample n_{tip}, n_{Sample}. As the *LDOS* often correlates to the presence or absence of atoms, the initial and most standard application of *STM* is the mapping of the surface topography. In practice, to obtain images of the surface topography, one scans across the surface in the so-called **constant current (CC)-mode**. In this mode, the tunnel current is kept constant by changing the distance between the sample and the tip accordingly. The corresponding backwards/forwards movement of the piezo-crystal, hosting the tip, are recorded and represent a map of the surface topography.

STM Spectroscopy, in contrast, requires measuring characteristic current-voltage ($I-V$) curves (or some derivative of it) as a way of recording the energy resolved *LDOS*, and is performed typically in **constant height (CH)-mode**, see *e.g.* Feenstra *et al.* [3], where the tip height is kept fixed with respect to the scanner head. A fixed tip height is maintained by breaking the feedback loop via a sample-and-hold amplifier (see section 3.1.1), which is controlled by a commercial software used in the present experiment. Operating the *STM* in *CH* mode also allows for a better normalization of the differential conductance dI/dV to the total conductance I-V, Feenstra *et al.* [3]. The *CH*-mode plays a central role in *NFESEM* and adds novelty to the instrument.

One key result to emphasize, because of its relevance for understanding *NFESEM*, is the lateral spatial resolution of a *STM*-like instrument. A careful theoretical analysis of the tunneling process between the tip and the sample [4, 5] reveals that the objects separated by a distance greater than:

$$\Delta S \simeq \sqrt{\frac{2\left(r_{tip}+d\right)}{\kappa}}, \tag{2.9}$$

are identifiable; where r_{tip} is the tip radius, d is the distance between the tip and the sample, and the inverse wavelength $\kappa = \left(2m_e\phi\right)^{1/2}/\hbar$ is defined by the effective barrier height ϕ between the tip and the sample. This expression and numerous experimental results show that atomic lateral spatial resolution can be achieved by imaging with a *STM*.

2.2. Field emission theory

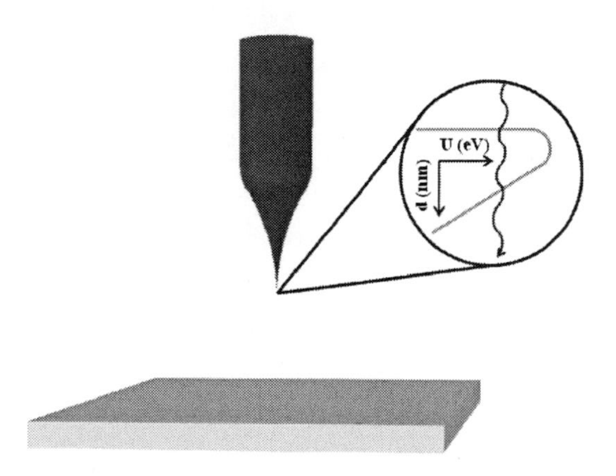

Figure 2.4.: Schematic of a needle electrode ejecting a field-emitted electron, through a potential barrier, towards a metallic surface.

Field emission is a phenomenon that occurs by submerging the surface of a condensed material, either solid or liquid, in high electrostatic fields ($> 1\,{}^{\mathrm{V}}/_{\mathrm{nm}}$) in order to expel electrons or ions into vacuum. The process entails the tunneling of electrons or ions through a deformed potential barrier at the surface (see Fig. 2.4 zoom). This is to be distinguished from thermionic emission or photoemission, where electrons are excited over the potential barrier. Therefore the surface potential is extremely important, as it has a profound influence on field emission properties. In fact, one of the main applications of field emission is the analysis of surface phenomena, *e.g.* the work function, which must be distinguished from the barrier height in *STM* [6]. Field ion emission closely resembles field electron emission (*FE*); however it occurs at even higher electric fields ($20 - 50\,{}^{\mathrm{V}}/_{\mathrm{nm}}$).

This section will only focus on field electron emission from metallic surfaces into vacuum. Field electron emission was one the first physical effects to be explained as due to quantum mechanical tunneling. The theoretical explanation was given by Fowler and Nordheim (*F-N*) in 1928 [7]. Versions of their famous equation are widely used, though many of these are only approximate solutions to the physical problem. *F-N* tunneling theory will be discussed in the following section, with special attention paid to its limitations.

2.2.1. Analysis of an exact triangular barrier

The purpose of the original paper on *Electron Emission in Intense Electric Fields* by R. H. Fowler and L. Nordheim [7] was to explain the current-(applied) voltage ($I - V_a$) characteristic for cold field emission (*CFE*) from metals, in particular the empirical law $\ln\{i\} \sim 1/V_a$ discovered by Millikan and Lauritsen (*M-L*) [8]. In some respects the *F-N* results resembled that of one derived from Oppenheimer's study of hydrogen in an external electric field [9], but the *F-N* findings are based on Sommerfeld's electron theory for metals and the exact solution of the Schrödinger equation. *F-N* reached the conclusion that at zero temperature, the expression for *CFE* current density J would have the form:

$$J = CF^2 e^{\alpha/F}; \tag{2.10}$$

where F is the field strength and C and α are constants, for which they gave explicit expressions (see below). This formula contains an F^2 term, which was not present in the *M-L* empirical formula. However, this difference was rightfully not considered important, given the electrical measurement technology, as it existed in 1928.

Although the *JWKB* approximation can, in principle, be applied to a triangular barrier, *F-N* chose to solve the Schrödinger equation exactly for the exact triangular barrier, depicted in Fig. 2.5a.

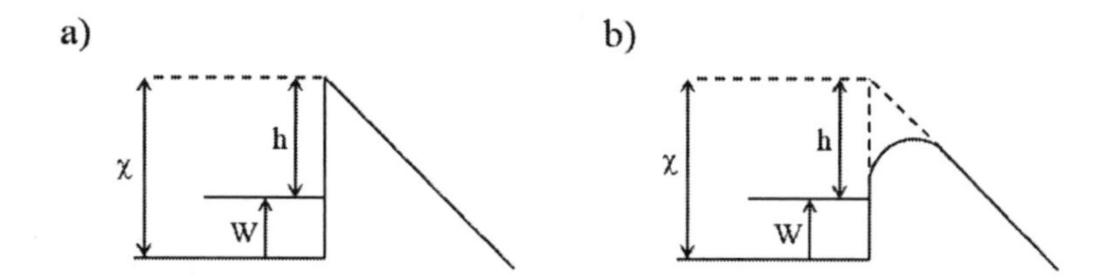

Figure 2.5.: Schematic of the potential barrier altered by an external electrostatic field without (a) and with (b) the distortion due to the image potential.

In Fig. 2.5 energies are measured relative to the bottom of the potential well. The y-axis denotes the energy component in the direction normal to the emitter surface. The "inner potential energy" (inner *PE*) χ is the total height of the *PE* step encountered. W is the component of electron kinetic energy in the direction normal to the emitter surface (this is sometimes called the "forwards kinetic energy").

For an electron with forwards energy W, the exactly triangular *PE* barrier is defined for $x \geqq 0$, by:

$$M(x) = (\chi - W) - eFx = h - eFx, \tag{2.11}$$

where the "zero-field barrier height" (h) is defined by:

$$h = \chi - W. \tag{2.12}$$

The Schrödinger equation must now be solved in two regions, in order to determine the emission through the potential barrier:

$$\left[\frac{d^2}{dx^2} + c^2\left(-h + eFx\right)\right]\Psi(x) = 0 \; (x > 0), \tag{2.13}$$

$$\left[\frac{d^2}{dx^2} + c^2 W\right] \Psi(x) \ (x < 0), \tag{2.14}$$

where $c^2 = 2m_e/\hbar^2$. The boundary condition is that Ψ and $d\Psi/dx$ are continuous at $x = 0$, *i.e.*, at the surface of the metal. Therefore, the most imperative equation is eq. 2.13, and the solution can be expressed as a Bessel function: $\Psi(x) = \sqrt{y} J_{\pm\frac{1}{3}}\left(\frac{2}{3}y^{3/2}\right) \ (x > 0)$ where:

$$y = \left(-\frac{h}{eF} + x\right)\left(c^2 eF\right)^{1/3}. \tag{2.15}$$

Additionally, imposing that the solution describes a wave traveling to the right, yields a similar expression with second function of Hankel:

$$\Psi(x) = \sqrt{y} H^{(2)}_{\frac{1}{3}}\left(\frac{2}{3}y^{3/2}\right) \ (x > 0). \tag{2.16}$$

Eq. 2.14 can easily be solved, *F-N* assumed the form:

$$\Psi(x) = W^{-1/4}\left[a e^{-icx\sqrt{W}} + a' e^{-icx\sqrt{W}}\right], \tag{2.17}$$

where a and a' are constants.

The equations of continuity[2] can now be imposed; hence

$$a + a' = W^{1/4}\left(\frac{\chi}{eF}\right)^{1/2} H^{(2)}_{\frac{1}{3}}\left(e^{-\frac{3\pi i}{2}}Q\right) \ \textit{and} \tag{2.18}$$

$$-a + a' = \frac{i}{cW^{1/4}}\left[\frac{1}{2}\left(\frac{h}{eF}\right)^{-1/2} + H^{(2)}_{\frac{1}{3}}\left(e^{-\frac{3\pi i}{2}}Q\right) + \frac{h}{eF}c\sqrt{eF}\frac{dH^{(2)}_{\frac{1}{3}}}{dQ}\left(e^{-\frac{3\pi i}{2}}Q\right)\right]; \tag{2.19}$$

where

$$Q = \frac{2}{3}c\frac{h^{3/2}}{eF}. \tag{2.20}$$

These two equations can be used to determine the fraction of electrons with energy W that penetrate the potential barrier.

The barrier penetration coefficient *D(F, W)* is given by $D(F,W) = \left[|a|^2 - |a'|^2\right]/|a|^2$. So from the equations above *F-N* obtain:

[2] There are a number of intermediate steps, which will not be restated; however it can be found in the Fowler-Nordheim article [7].

$$D(F,W) = \frac{4\beta\left(\frac{h}{eF}\right)^{3/2}\sqrt{eF}}{\left[W^{1/4}\left(\frac{\chi-W}{eF}\right)+\frac{\beta h}{W^{1/4}\sqrt{eF}}\right]^2+\frac{1}{c^2W^{1/4}}\left[\frac{1}{2}\left(\frac{h}{eF}\right)^{-1/2}+c\alpha\left(\frac{h}{\sqrt{eF}}\right)\right]^2}; \tag{2.21}$$

where α and β are parameters discussed by *F-N* [7]. They go on to show that the equation above can be reduced to an equation that, when written using modern notation, has the form:

$$D(F,W) = P^{FN}\exp\left[-bh^{3/2}/F\right], \tag{2.22}$$

where b is the so-called "second *F-N* constant" and is given by $b = (4/3)\,(2m_e)^{1/2}/e\hbar \approx 6.830890\ \mathrm{e}V^{3/2}\ \mathrm{V}\ nm^{-1}$, and the *F-N* tunneling prefactor P^{FN} is given by:

$$P^{FN} = \left(4W^{1/2}h^{1/2}\right)/\chi. \tag{2.23}$$

Finally, *F-N* sum over all occupied electron states by using Nordheim's supply function *N(W,T)* [10] for the number of electrons per unit area per second of energy within dW striking the surface potential barrier. This yields the integral:

$$J(F,T) = e\int_0^\infty N(W,T)\,D(F,W)\,dW. \tag{2.24}$$

In the zero-temperature limit, this integral can be evaluated to give the "original" *F-N*-type equation, which in modern form can be written:

$$J = P_F^{FN}a\varphi^{-1}F^2\exp\left(-b\varphi^{3/2}/F\right). \tag{2.25}$$

Here P_F^{FN} is given by $\left(4\mu^{1/2}\varphi^{1/2}\right)/\chi$, where μ is the Fermi energy, *i.e.* the kinetic energy of an electron at the Fermi level, and a is the "first *F-N* constant", given by $a = e^3/8\pi h_P \approx 1.541434\cdot 10^{-6}\mathrm{A}\ \mathrm{eV}\ V^{-2}$, where h_P is Planck's constant. In the literature P_F^{FN} is often set equal to unity, giving the equation:

$$J = a\varphi^{-1}F^2\exp\left(-b\varphi^{3/2}/F\right). \tag{2.26}$$

This is the so-called "elementary" *F-N*-type equation, and has been very widely used.

2.2.2. Analysis of the Schottky-Nordheim barrier

Assuming an exact triangular barrier, as in Fig. 2.5a, allows a mathematically exact analysis, but the assumption is not physically realistic. A more reasonable approximation of the potential incorporates a classical image potential energy, which is depicted in Fig. 2.5b.

For an electron with forwards energy W, this potential is described by:

$$M(x) = h - eFx - \frac{e^2}{16\pi\varepsilon_0 x}. \tag{2.27}$$

This barrier is sometimes called the Schottky-Nordheim (*SN*) barrier. Consequently eq.2.13 becomes:

$$\left[\frac{d^2}{dx^2} + c^2\left(-h + eFx + \frac{e^2}{16\pi\varepsilon_0 x}\right)\right]\Psi(x) = 0\ (x > 0). \tag{2.28}$$

This differential equation has no analytical solutions in terms of the usual functions of mathematical physics. Consequently, it is in practice always solved using a *JWKB*-type approximation. Mathematical errors occurred in early work, and the first researchers to obtain a reasonably correct solution were Murphy and Good [11]. For *CFE*, their solution can be written in the form:

$$J = t_F^{-2} a\varphi^{-1} F^2 \exp\left(-v_F b\varphi^{3/2}/F\right), \tag{2.29}$$

where v_F is a particular value of a mathematical function v known either as the "principal field emission elliptic function" or (more recently [12]) as the "principal Schottky-Nordheim barrier function", and t_F is a particular value of a function t related to the function v. The particular values v_F and t_F correspond to a tunneling barrier of zero-field height h equal to the local work function φ. Eq. 2.29 is sometimes called the "standard" *F-N*-type equation.

Murphy and Good [11] obtained an analytical expression for v in terms of complete elliptical integrals of the first and second kind. Recently, Forbes [12] has found the following simple approximate expression for v_F :

$$v_F \approx 1 - \frac{F}{F_\varphi} + \frac{F}{6F_\varphi} ln\left(\frac{F}{F_\varphi}\right); \tag{2.30}$$

where F_φ is the field necessary to reduce a barrier of initial height φ to zero.

A number of approximations were used to arrive at the original *F-N*-type equation, and these significantly restrict its applicability and/or accuracy. For instance, it has been assumed that: the atomic structure can be ignored and Sommerfeld's free electron model can be used; the electron distribution is in thermodynamic equilibrium and obeys Fermi-Dirac statistics; the temperature of the system is at zero; the surface of the emitter is planar, with a constant uniform work function; there is a uniform electric field outside; a classical image potential can be applied; and the barrier penetration coefficient can be determined using the *JWKB* approximation. These points will addressed in the following section with the emphasis of applying *F-N* theory to field emitters, specifically field emitters with a radius of less than 10 nm.

2.2.3. The tunneling prefactor

It will be noted that the original *F-N*-type equation contains a tunneling prefactor P_F^{FN}, but that the standard *F-N*-type equation does not. Forbes [13] has recently argued that all *F-N*-type equations should, in principle, include a tunneling prefactor. In general, a *JWKB*-type approximation [14] is used to determine the barrier penetration coefficient (D). For the strong barriers considered in *F-N*-type theory, D should have the form[15]:

$$D = Pe^{-G}. \tag{2.31}$$

Usually the prefactor P can be approximated to unity [12, 13].

Note that, in contrast to 'strong' barriers, 'weak' barriers, *i.e.* $\exp -G \sim 1$, require a more complete approximation, for instance the Fröman and Fröman exact calculation of D [14]. In the light of Forbes's [13] formulation, using a one-dimensional treatment, the exact solution can be expressed as:

$$D = Pe^{-G} / \left(1 + Pe^{-G}\right). \tag{2.32}$$

Equations that include this expression for D are not considered to be F-N-type equations.

2.2.4. The physically complete Fowler-Nordheim-type equation

Forbes has also suggested that, in principle, *CFE* current density should be given by a so-called "physically complete" F-N-type equation that has the form:

$$J = (\lambda_Z a \varphi^{-1} F^2) P_F \exp\left[-\nu_F b \varphi^{3/2} / F\right], \tag{2.33}$$

where: ν_F ("nu_F") is a correction factor associated with barrier shape; P_F is a tunneling prefactor; and λ_Z is a correction factor associated with the effective electron supply - this takes into account effects due to integration over states, electronic structure, and temperature. Obviously, this equation contains three correction factors: ν_F, P_F, and λ_Z. The precise forms of these correction factors depend on the physical features of the emission situation under analysis. For example, they might depend on the geometry of the emitter and/or on the chemical nature of its surface.

2.2.5. The need for auxiliary equations

To relate a measured $I - V_a$ characteristic to a theoretical discussion, in terms of current J and field F, it is necessary to introduce two axillary equations. First, a reference point "0" is chosen on the emitter surface. Often this is the point at which the current density is a maximum. Denote the field at this reference point by F_0 and the current density at this reference point by J_0. This field may be related to the applied voltage V_a by the equation:

$$F_0 = \beta_0 V_a, \tag{2.34}$$

where β is the voltage-to-barrier-field conversion factor, and β_0 is the value of β at the reference point. Values for β_0 have to be obtained by electrostatic analysis of the emitter and system geometry.

With real emitters, the emission current density J varies with position across the surface, and the total emission current I has to be determined (at least in principle) by integration. The result of this integration is written in the form:

$$I = \int J\, dA = A_n J_0, \tag{2.35}$$

where A_n is called the "notional emission area." In reality this parameter A_n may be a function of the emitter geometry, the applied voltage, temperature, and possibly other factors; however it is often approximated as being constant.

2.2.6. The relationship between voltage and field

Two limiting forms for the conversion factor β are of interest here. In the context of a parallel plate arrangement β is given by:

$$\beta = 1/w, \tag{2.36}$$

where w is the plate separation.

For a sphere, β would be given by: $\beta = 1/r_{sphere}$. For a field emitter with apex radius r_{tip}, it is conventional to write β_0 in the form:

$$\beta = \frac{1}{k_f \cdot r_{tip}}, \tag{2.37}$$

where k_f is called the " shape factor" or the "field reduction factor." When the emitter is distant from its surroundings, it is often assumed that k_f has a value in the range of 5 - 8. The value 5 is often used as an easy rough approximation.

Obviously eq. 2.37 shows that β_0, and hence F_0, depends very strongly on the apex radius, particularly when this is small.

As a conventionally shaped emitter approaches (from a large separation) towards a planar surface, the most useful formula for β changes from eq. 2.37 to eq. 2.36. However the manner in which this change takes place depends on the detailed geometry of the emitter, and is usually not well-known.

Note that in the parallel plate arrangement, if there is a small hemispherical protrusion on one of the plates then the field at the apex of this hemispherical protrusion is equal to three times the field between the plates at large distances from the protrusion.

2.2.7. The "Sphere-on-orthogonal-cone" model

An early examination of geometric effects on field emission was performed by Dyke *et al.* [16], who considered the so-called "sphere-on-orthogonal-cone" model for the shape of a field emitter. This cathode model uses the electrostatic potential distribution:

$$V = P_n(\cos\theta)\frac{V_c}{R^n}\left(r^n - a^{2n+1}r^{-n-1}\right); \tag{2.38}$$

where P_n is Legendre function, with n chosen so that θ vanishes at the exterior half angle of the cone, V_c is the electrostatic potential difference between the "core" of the cathode model and the anode, R is the anode distance, a is the sphere radius, r and θ are ordinary polar coordinates. Setting $V = V_c$, they find that the shape of the anode is given by:

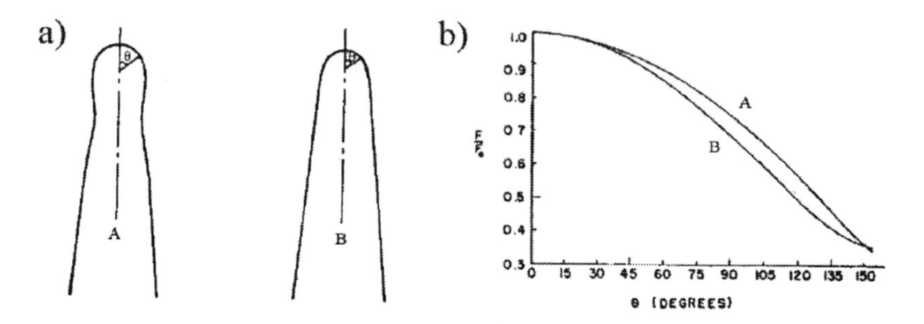

Figure 2.6.: Schematic of (a) emitter profiles with a neck constriction on the left and without a constriction on the right and (b) the corresponding field distribution as a function of the polar angle [16].

$$r^n = \frac{R^n}{P_n \cos\theta}; \tag{2.39}$$

hence it describes a surface resembling a paraboloid.

The emitter surface is modeled by one of the equipotential surfaces described by eq. 2.38. Let this equipotential surface correspond to a potential difference V_a between this surface and the anode. Dyke *et al.* show that if we take the reference point discussed earlier as the "apex" of this equipotential surface, then the conversion factor β_0 (as given by the *SOC* model) is:

$$\beta_0 = \frac{F_0}{V_a} = \left(\frac{V_c}{V_a}\right)\left[n + (n+1)\left(\frac{a}{r_0}\right)^{2n+1}\right]\left(\frac{r_0^{n-1}}{R^n}\right), \tag{2.40}$$

where F_0 (as before) is the field strength at the apex and r_0 is the value of r at $\theta = 0°$ on the surface which approximates the emitter.

Dyke *et al.* also give an equation for the ratio β/β_0, and were able to predict the polar distribution of the field strength and of the corresponding current density. For a given geometry, (see Fig. 2.6a), the dependence of the field strength upon polar angle is shown in Fig. 2.6b. The current density falls off somewhat more rapidly, because of its exponential dependence on *1/F*.

2.2.8. The electrostatics of emitters close to surfaces

The results for the *SOC* model (just quoted) apply when the emitter is well-distant from its surroundings. When an emitter is relatively close to a surface, as it is in *NFESEM*, the electrostatics of the whole configuration must be analyzed consistently. There seem to be no suitable analytical treatments, and there appear to be no published numerical simulations that correspond exactly to the geometry of the present experimental arrangements. For very small separations between a field emitter and a planar surface, Mesa *et al.* have carried out numerical simulations [17]. These show that the shape factor k_f varies approximately linearly with tip-sample separation, although they also find that the relationship is not exactly linear. The gap distance dependence on the applied voltage, V_a, for a small hemispherical protrusion ($r_{protrusion} = 0.05r_{tip}$), is shown in Fig. 2.7. It is necessary for *NFESEM* emitters to have such asperities on the tip, in order to generate high electric fields at the apex.

2.2.9. Field emission microscopy

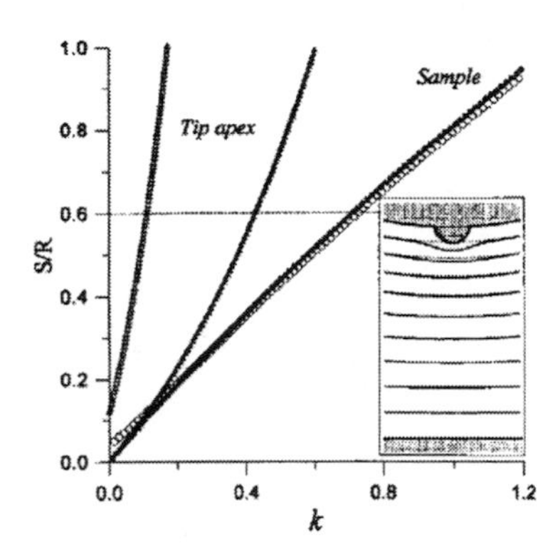

Figure 2.7.: Numerical simulation of the tip-sample separation normalized by the emitter radius (S/R), where R = r_{tip}, vs. $k_f = V_a/F \cdot r_{tip}$. The solid circles correspond to a "smooth" emitter. Open symbols represent the changes in k_f due to the appearance of a small hemispherical protrusion ($r_{protrusion} = 0.05r_{tip}$) on the tip apex. The inset shows the influence of an asperity in the equipotential lines for a distance of $0.6r_{tip}$ [17].

The field emission microscope (*FEM*) (also called a "field electron microscope") uses the projection of field-emitted electrons that are accelerated towards a conducting fluorescent screen to generate an image. Normally, the sample (or field emitter) is a metallic tip of high curvature; hence high surface fields can be generated with moderate applied voltages. Ordinarily, the *FEM* uses a purely *CFE* process. Field-emitted electrons from a region of size ξ on a tip with a radius r_{tip}, generate a spot of size Ξ, when projected onto a plane at a distance X from the sample [18]. Ξ is given by:

$$\Xi = \frac{\xi \cdot X}{r_{tip} \cdot \beta_{comp}}, \tag{2.41}$$

where β_{comp} is the so-called "compression factor." β_{comp} is typically around 1.5 when $X/r_{tip} > -10$.

I-V characteristics can be readily determined with a *FEM*.

2.2.10. Analysis of experimental data

When the auxiliary equations are used to convert J and F, the physically complete *F-N*-type equation becomes (in its $I - V_a$ form):

$$I = (A_n \lambda_Z a \varphi^{-1} \beta^2 V_a^2) P_F \exp\left[-\nu_F b \varphi^{3/2} / \beta V_a\right]. \tag{2.42}$$

In so-called *F-N* coordinates, this becomes:

$$\ln\left\{I/V_a^2\right\} = \ln\left\{A_n\lambda_Z a\varphi^{-1}\beta^2 P_F\right\} - \nu_F b\varphi^{3/2}/\beta V_a. \quad (2.43)$$

A diagram in which $\ln\{I/V_a^2\}$is plotted against $1/V_a$ is called a "Fowler-Nordheim plot". It is conventional to plot experimental *CFE* data as a *F-N* plot. Experimental data plots of this kind are often (but not always) straight lines. Fitting a straight line to this experimental data is equivalent to finding the equation for the *tangent* to the curve represented by eq. 2.43, at some point on the curve. It can be shown that the slope of this tangent is theoretically predicted to be:

$$m = -\sigma b\varphi^{3/2}/\beta, \quad (2.44)$$

where σ is a generalized slope correction factor. For the Schottky-Nordheim (*SN*) barrier, σ is given by a mathematical function s, values of which are tabulated in the literature. Typically, s is of the order 0.9 - 0.95. Hence, a *SN* barrier predicted slope of a *F-N* plot can be written in the form:

$$m = -sb\varphi^{3/2}k_f \cdot r_{tip}. \quad (2.45)$$

2.2.11. Correction for sharp emitters: Cutler-Edgcombe correction

Until now, only emitters with a sufficiently large radius of curvature have been discussed. P. Cutler *et al.* [19] have explored the breakdown of standard *F-N* theory for emitters with a high radius of curvature and have shown that small-radii emitters exhibit deviations from linear *F-N* curve behavior. Because their models (in effect) assume that the emitter is relatively close to a planar anode, their work is particularly relevant to the present dissertation. They have clearly shown that the standard *F-N*-type equation does not apply to situations where the emitter is sharply curved. Rather, the correction factor ν_F, in the physically complete *F-N*-type equation must be replaced by something more general than the mathematical function v. Modifications to the pre-exponential will also be necessary. It is important to mention that deviations from *F-N* plot linearity, for sharp emitters, are related to neither space charge effects nor surface roughness [6, 20]. Usually the current density is lower than the threshold for space charge effects [21, 22], which is generally around $J > 10^6\,\mathrm{A/cm^2}$.[3]

Even with classically flat emitters, there are effects that would cause *F-N* plots to be slightly curved. However with sharp emitters, the main effect that causes non-linearities in the *F-N* plot is the fall-off in the strength of the electrostatic field with distance from the emitter surface. For this reason, revisions of the *F-N* equation are required for nanometer emitters, which must include an asymmetric potential barrier that varies with the tip radius and the polar angle.

P. Cutler's [19] method involves numerical calculations for specified emitter geometries, *e.g.* hyperboloid and cone, which influence the barrier shape. The resultant *F-N* curves were fitted to the following equation:

$$J = AV_a^2 \exp\left(-B/V_a - C/V_a^2\right); \quad (2.46)$$

[3]Please refer to chapter 2 of [21] for an involved discussion of the space charge effect.

where A, B, and C are constants that depend on material and geometrical properties.

G. Fursey has also prescribed a barrier model. In effect, he expands on the motive energy in eq. 2.28 to include the field fall-off. This leads to the expression:

$$M(x) = h - eFx\left(\frac{r_{tip}}{x + r_{tip}}\right) - \frac{e^2}{16\pi\varepsilon_0 x}. \tag{2.47}$$

It follows that defining a suitable potential barrier may enable the determination of microscopic information about the emitter from $I - V_a$ characteristics [23, 24]. Recently, a *FE* curved-surface theory [24] was developed to explore the emission properties of a carbon nanotube (*CNT*). This demonstrated the theory's ability to deduce the work function of the surface, along with the emitter radius, the surface field, the effective solid angle of emission and the supply factor from measured characteristics. C. Edgcombe [24] has derived a direct relationship between the curvature of a *F-N* plot and a function describing a hemispherical barrier, as it is varied along the emitter surface:

$$\frac{\partial S/\partial V_a^{-1}}{S \cdot V_a} = x\frac{\partial^2 f_1/\partial x^2}{\partial f_1/\partial x}; \tag{2.48}$$

where S is the rate of change of the exponent of the measured current dependence with V_a^{-1}, x is the ratio of the minimum barrier thickness to the emitter radius of curvature, and f_1 is the barrier integral. The curvature of the *F-N* plot is determined via a quadratic fit of the form:

$$\ln\left(I_{FE}/V_a^2\right) = AV_a^{-2} + BV_a^{-1} + C; \tag{2.49}$$

where A, B, and C are fitting parameters (different from the ones used above). Note that this is the same dependence observed by Cutler *et al.* [19]; hence this method is appropriate for emitters that have a radius of curvature equal to (or less than) the barrier width. The fitting parameters evaluated at an arbitrary V_a^{-1} are used to estimate a value for the emitter radius using the following formula:

$$\sqrt{\varphi} \cdot r_{tip} = (c_2 e x)^{-1}\frac{-SV_a^{-1}}{\partial f_1/\partial x}, \tag{2.50}$$

here $c_2 = \left(\frac{4\sqrt{2m_e}}{3e\hbar}\right)$.

This generated a reasonable estimation of the associated field emitter properties using only the curvature of the *F-N* plot and the energy distribution of the field-emitted electrons. The work function and subsequently the emitter radius can be determined more precisely using the parameter d_0, which is a parameter that relates the rate of change of the barrier penetration coefficient D , usually the "decay width", to electron energy W near the Fermi level E_f [25]. A similar procedure was also applied to emitters used in the present microscope and will be discussed later.

2.3. Modeling the interaction volume with the primary electron beam

In order to interpret features generated in *NFESEM*, it is necessary to formulate a model of describing the interaction of the primary field-emitted electrons with the specimen. The primary electrons undergoes elastic and inelastic scattering processes, within the specimen, resulting in zig-zag trajectories until the electrons gradually decelerate via energy loss and come to rest or leave the specimen as backscattered electrons. Thus, the angular, spatial, and energy distributions of the electron trajectories is a complex process involving multiple scattering. The diffusion of electrons in a solid is typically treated within transport equations or the Monte Carlo method.

These methods take into account:

a. the elastic processes, described by the famous Rutherford differential elastic cross section:[4]

$$\frac{d\sigma_{el}}{d\Omega} = \frac{e^4 Z^2}{16(4\pi\epsilon_0 E^2)} \frac{1}{[(sin^2(\theta/2))]^2}; \tag{2.51}$$

where $d\Omega = 2\pi sin\theta d\theta$ is the element of the solid angle into which the electron energy E is scattered at an angle θ from its incident direction, e is the electron charge, Z is the atomic number of the scattering atom, and ϵ_0 is the electric constant. When atoms of heavy elements are present in the specimen, spin-orbit coupling is important and the Rutherford elastic cross section is corrected to a Mott scattering cross section [27].

b. Inelastic processes, associated with an energy loss of the primary electrons and a transfer of this energy to other electrons or other types of excitations (plasmons, phonons, *etc.*).

c. Multiple scattering, due to the fact that electrons can make many elastic and inelastic scattering processes when diffusing within a solid.

Qualitatively, the main result of these processes is that the energy spectrum, of emitted electrons, is characterized by a large number of secondary electrons (SE)s with energies in the range $0 - 15$ eV above the vacuum level. This is followed by a weak background onto which characteristic, typically weak, losses are superposed, *e.g.* Auger electrons, plasmons and core level excitations. Finally, a substantial elastic peak is observed, formed by those electrons which are backscattered without loss of energy. The energy spectrum is extremely complex and depends on the primary electron energy used.

At moderate and high primary energies ($\geq 1keV$) reliable Monte Carlo simulations have been developed, such as the CASINO software [28], which allow to compute *e.g.* trajectories of electrons entering the solid. Fig. 2.8 is a CASINO Monte Carlo simulation using typical *NFESEM* values, *i.e.* a probe diameter of 1 nm with a primary beam energies ranging between 30 - 90 eV. In addition the following physical models were applied for the associated calculation: total and partial cross section - Mott by interpolation, effective section ionization - Casnati [30], ionization potential and ${}^{dE}/_{dS}$- Joy and Luo (1989) [31], random number generator - Press *et al.* (1992) [32], and directing cosin - Soum *et al.* (1979) [33]. The result of this tentative simulation show that:

1. the penetration depth of the electrons is limited to the top surface layers, *i.e.* experiments using these primary energies are extremely surface sensitive

[4] See [26] for a complete derivation of the Rutherford differential cross section

2. the interaction volume is not much larger than the original beam diameter.

Even at 90 eV the interaction volume diameter is slightly over 2 nm, using a primary beam diameter of 1 nm. In accordance with these preliminary simulations, the spatial resolution of *NFESEM* will be approximately equal to the impinging beam width. Therefore, the important aspects concerning this last quantity will be discussed in the following section.

It is important to note that the CASINO simulation is not recommended for low primary beam energies ($\leqq$ 50 eV) [28], since the scattering mechanism is assumed to be constant;[5] thus neglecting surface plasmon decay into *SE*s and the proportionality of electron scattering cross section to empty *d*-states [35].[6]

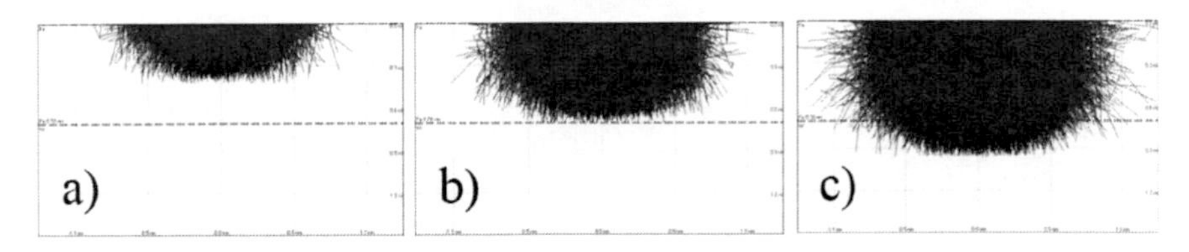

Figure 2.8.: Monte Carlo simulation of the electron trajectories interacting with 7Å of iron on a tungsten substrate at a primary beam energy of (a) 30 eV, (b) 60 eV, and (c) 90 eV using the CASINO program [28]. The beam width is 1 nm, the number of primary electrons simulated is 10^4, secondary electrons and backscattered electrons are represented in blue and red respectively. The dimension of the simulation is 3 x 2 nm.

[5] Calculated and measured electron inelastic mean free path show an up-turn at lower energies, due to an inability to excite plasmons [34].

[6] A large, international collaboration directed towards a greater understanding of low energy (1 - 200 eV) electron interactions is currently underway.

2.4. Resolution capabilities of the NFESEM

In Scanning Electron Microscopy (*SEM*), including *NFESEM*, the wavelength of the electrons – determining the diffraction limit for the lateral spatial resolution – is on the order of 0.1 nm or less. Accordingly, the lateral spatial resolution is effectively defined by the actual lateral length scale onto which the electron beam can be focused. Focusing the electron beam, technically, is the difficult part of any *SEM*-type instrument. In *SEM* with a remote source, this problem is solved by a sophisticated sequence of electrostatic and magnetic electron lenses. In lensless *NFESEM*, one has to rely on the geometrical and physical properties of the field-emitted beam.

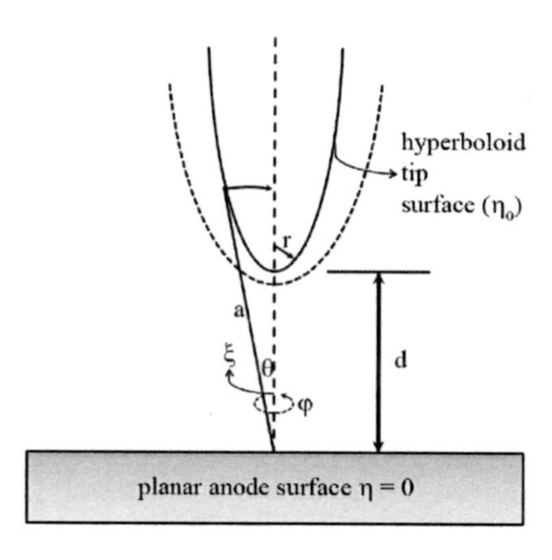

Figure 2.9.: Schematic of the tip-sample geometry in a prolate spheroidal coordinate system.

J.J. Sáenz [17, 36] has considered a model consisting of a hyperboloidal emitter cathode and a planar anode, as illustrated in 2.9, in the "near field" emission regime and computed the lateral spatial resolution to be:

$$\Delta x \approx 0.7 \cdot [(r + d) \cdot d]^{1/2} . \tag{2.52}$$

As a result the lateral resolution dependence on the emitter radius and tip-sample separation indicate that atomic lateral resolution is not feasible under the conditions investigated by Sáenz. At distances used in the *NFESEM* experiment, typically d is of the order of a few tens of nanometers, Sáenz estimates a lateral resolution of the order of d.

In the same paper Sáenz also predicts, in the *NFESEM* regime, a possible vertical resolution of the order of less than 0.2 nm and demonstrates this performance by experimentally detecting a step edge on a graphite surface [36]. In Chapter 4, we will discuss the lateral and the vertical resolution in light of the *NFESEM* results [38, 39]. Note that all of the *NFESEM* experiments were performed at room temperature; therefore thermal effects on the emitter were not considered. Although it has been determined that the electron beam is partially coherent at room temperature, cooling the emitter will enhance coherence [37]. This can possibly increase *NFESEM* resolution capabilities.

3. Experiment

In the previous chapter, the fundamental principles of the primary beam generation, the interaction of this beam with the specimen and the theoretically expected spatial resolution have been discussed. This chapter illustrates the experimental realization of this instrument, including tip and sample preparation and characterization techniques for both. Recall that the main aim of this instrument is the realization of some kind of surface topography image due to the exposure of a primary beam of electrons, as it is rastered along the sample surface. This will be achieved by two distinct, although related, experiments: measuring the field emission (*FE*) current while scanning and detecting the secondary electrons (*SE*)s generated when the electron beam impinges on the surface.

3.1. Experimental Set-up

R.D. Young *et al.* [40] were the first to engineer a localized field emitter tip as a means of generating a primary electron beam. In the experimental set-up proposed by R.D. Young, see Fig. 3.1, a tip is brought into close proximity to the sample surface until the desired *FE* current is achieved. The tip is then rastered along the surface at a fixed *FE* current, which is maintained by a feedback controller that adjusts the vertical z-piezo voltage accordingly. The applied voltage is much larger than the work function of the emitting material, which results in a current-(applied) voltage ($I - V_a$) relation consistent with conventional Fowler-Nordheim *FE* theory [7]. This, nearly mono-energetic, primary beam of electrons impinges on the surface at a well-defined area, without the aid of focusing optics, and undergoes elastic as well as inelastic collisions penetrating the outermost layer(s) of the surface (less than 1 nm). Subsequently, the ejected *SE*s from the excitations are collected and detected using an electron multiplier. The variations in the electron intensity from the electron multiplier signal, as a function of scanning position, represent the surface topography. Although this experiment by R.D. Young *et al.* is conceptually similar to a scanning tunneling microscope (*STM*), it however generates images

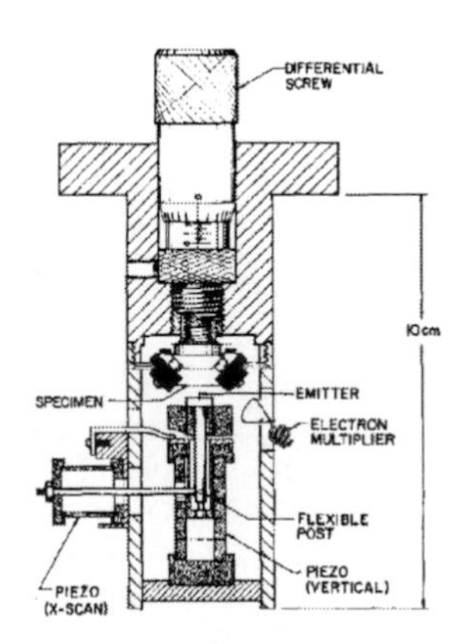

Figure 3.1.: Schematic drawing of the topografiner [40].

similar to a scanning electron microscope (*SEM*), without the need of a "remote" electron gun.

More than a decade later, H.-W. Fink achieved 3 nm lateral resolution on a polycrystalline gold surface [41, 42], using a (111)-oriented tungsten (*W*)-tip as a source [41, 42, 43]. The primary beam of electrons was extracted from the *STM* tip via an aperture placed in between the tip and the sample. These tips required field evaporation, *e.g.* with field ion microscopy (*FIM*), to atomically engineer the apex. The tip was rastered near the surface using constant current (*CC*) mode at an emission current of 0.1 nA and primary beam energy of 15 eV. In addition, the primary electron beam diameter was reduced via an aperture placed in between the tip and sample [42].

P.N. First *et al.* [44] performed similar measurements as described by H.-W. Fink; in addition, spin analyzers were employed to detect the surface magnetization vector. The topography, the *SE* intensity, and the magnetization of the sample were simultaneously measured with a lateral resolution of about 50 nm. The tip was rastered using *CC* mode at a nominal distance of 100 nm to the sample, dynamically adjusted by a correspondingly controlled z-piezo using the "topografiner mode". Voltages ranging from 32 – 45 V were applied between the tip and sample. The resulting topographic and *SE* intensity image exhibited the converse contrast, *i.e.* protrusions in the topographic image appeared as depressions in the *SE* intensity image.

In an alternative configuration, a *STM* operating in *FE* mode was used to map the emission sites of a broad-area cathode, where the tip is held at a positive bias in order to extract field-emitted electrons from the sample; thus providing sample topography, surface potential distribution and conductivity, and emission intensity [45].

In a complementary experiment, R. Palmer *et al.* performed electron energy loss spectroscopy (*EELS*) using a single crystal (111)-oriented *W*-tip as a primary beam emitter [46, 47]. Here an energy analyzer was used to determine the chemical properties of the surface. The tip radius, as determined by *SEM*, was approximately 10 - 30 nm and obtained a lateral resolution of approximately 40 nm. Movement of the tip, in the vertical direction, was controlled by a feedback loop controlling the z-piezo, *i.e.* *CC* mode, to maintain a distance of less than 200 nm. The backscattered electrons (*BSE*)s entered the four grid retarding field analyzer through a grounded first grid. No direct relationship was drawn between the highest topographic area and the highest reflected electron count rate image; however angular resolved *EELS* was performed, where a distribution of *BSE*s and *SE*s was observed at angles ranging from $-0.8°$ to $18°$, with respect to the sample surface plane [48]. This shows that the majority of electrons were ejected parallel to the surface, but electrons of highest energy loss exhibited trajectories bent towards the optical axis, in the direction of the primary beam.

The present *NFESEM* system consists of a homemade modified Lyding-type *STM*[1] [49] and a modified *SE* detector (*SED*), as shown in Fig. 3.2a. *UHV* conditions are required to reduce surface contamination, which significantly alters the *SE* yield, as well as increase the primary beam stability. Therefore, the system was installed in a specially designed titanium *UHV* chamber with an inner aluminum coating, enabling the system to achieve a base pressure lower than $2 \cdot 10^{-11}$ mbar. Moreover, the *STM* is suspended by

[1] The distance along the tip axis and the distances parallel to the surface are obtained by reading the voltage applied to the piezo-crystals supporting the tip. A calibration of the applied voltage against lateral and vertical movements was performed by using *STM*-imaging of a Si (111) 7×7 reconstructed surface, for which both lateral and vertical spacings are well-known. The error, in measuring absolute distances, is less than 1 Å.

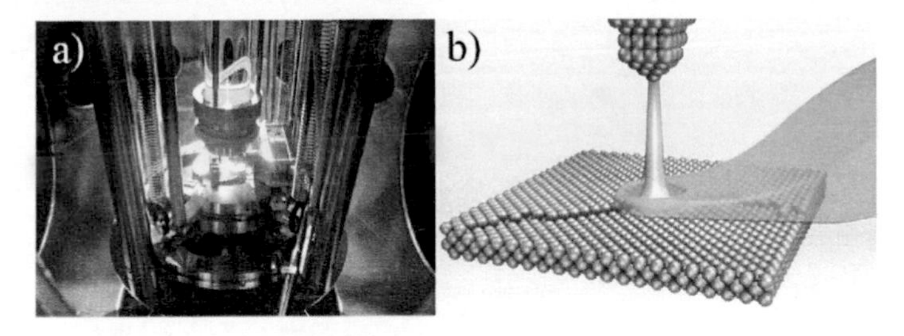

Figure 3.2.: (a) Actual picture and (b) artistic rendition of the *NFESEM*. In this instrument electrons are emitted from the sample surface after undergoing interactions with a primary beam of electrons field-emitted from a sharp tungsten tip, positioned very close to the surface via a piezoelectric device.

four stainless steel springs for high frequency damping and, additionally, has eddy current damping on the bottom side of the base plate. The geometry of the *STM* allows for the detectors to be mounted near the sample. Our *SED* is situated approximately 2 cm from the sample edge and aligned to collect electrons ejected parallel to the plane of the sample surface, see Fig. 3.2a, in accordance with the deflection of electrons by the strong electric field between the tip and the sample. Fig. 3.2b shows an artistic, atomic scale, scheme of the experimental configuration.

In *NFESEM*, a primary beam of electrons impinge on the sample surface at currents on the order of tens of nanoamperes. Even though higher currents may increase the resolution of the image, it may also contaminate the surface and/or induce adsorbate motion between the tip and the sample. The beam energy should be the minimum energy required to eject several electrons from the top-most layer of the surface, presumably less than 60 eV. Secondary and reflected electrons will be deflected by the strong electric field ($\sim 10 - 30\,{}^{\mathrm{V}}/_{\mathrm{nm}}$) between the tip and sample, which directs the escaped electrons parallel to the surface. The electron detector, mounted essentially parallel to the sample surface, should cover an area large enough to encompass the entire sample. An acceleration voltage may be applied to the scintillator disk or extraction optics; however it should not perturb the primary beam. Here the illustration shows a photo-multiplier tube (*PMT*) and a scintillator used for detecting the *SE*s. The extraction voltage used is determined by the minimum acceleration voltage required to excite measurable photons at the scintillator disk, located at the entrance.

The *SE*s are accelerated to a $YAlO_3$ single crystal scintillator, in the perovskite phase, (*YAP*) at energies up to 3 keV; where they are converted to photons that travel along a 45 cm Pyrex light guide to a Hamamatsu R268 *PMT*,[2] as depicted in Fig. 3.3. A gold-coated MACOR piece insulates the high voltage at the titanium ring and is at the same ground as the detector. Our *SED*, comprised of the scintillator and *PMT*, is mounted on a linear *UHV* manipulator with a travel distance of 10 cm. In general, the primary beam energy is on the order of tens of eV; hence the *BSE*s have low energies. Both *BSE*s and *SE*s are sampled in the current setup, due to a large detector surface area and high electron acceleration voltages. This *SED* is reminiscent of an Everhart and Thornley (*ET*)-type detector [50], with the exception of the Faraday cage preceding the scintillator. In general,

[2]The quantum efficiency of this *PMT* is 28% with an optimal photon wavelength of 390 nm.

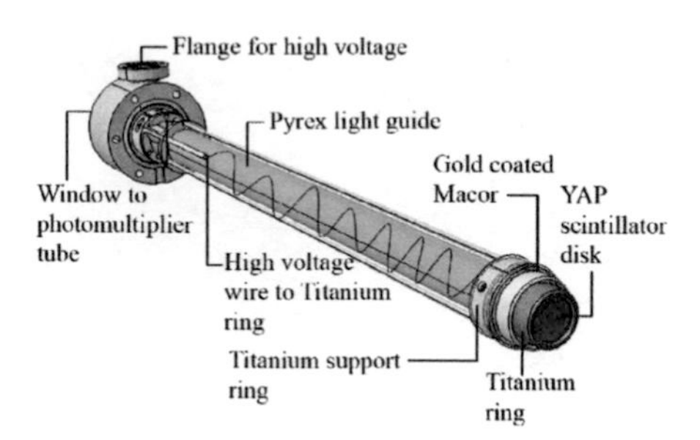

Figure 3.3.: Schematic diagram of the secondary electron detector.

the *ET* detectors are sensitive to low energy *SE*s, since the solid angle of collection is too small for *BSE*s; hence the measured *SE* current is:

$$I_{SE} = I_P f_{SE} \delta; \tag{3.1}$$

where I_P is the probe current, f_{SE} is the collection efficiency, and δ is the *SE* yield. One must also consider the effect of pressure on the I_{SE}, since sensitive detectors will exhibit a significant increase in the background signal as the pressure is increased from $2 \cdot 10^{-11}$ to $5 \cdot 10^{-11}$ mbar.

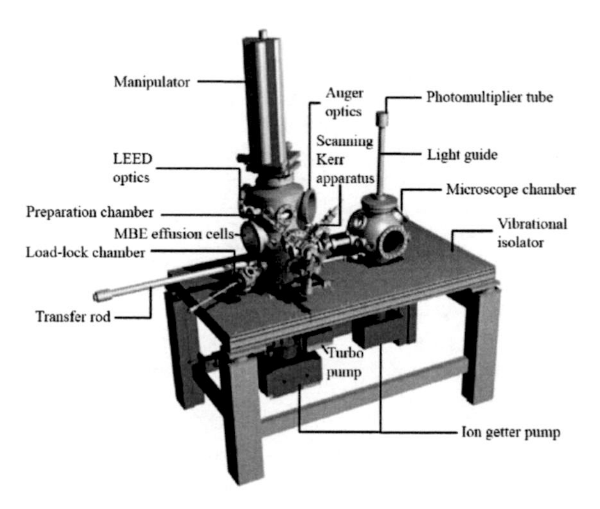

Figure 3.4.: The ultra-high vacuum system used to perform *NFESEM*. The sample preparation chamber is on the left side, and the microscope is located directly below the light guide on the right side.

The samples and tips were prepared and characterized in situ via an adjacent interlocked *UHV* chamber with a base pressure of less than $1 \cdot 10^{-10}$ mbar accommodated with low-energy electron diffraction (*LEED*) optics, optics for Auger electron spectroscopy (*AES*), sputter gun, mass spectrometer, molecular beam epitaxy capabilities, heating stage, and a sample/tip docking carousel, see Fig. 3.4. Sample preparation is extremely important for the *SE* yield, since the primary electron beam only penetrates a few atomic layers.

3.1.1. Electronics

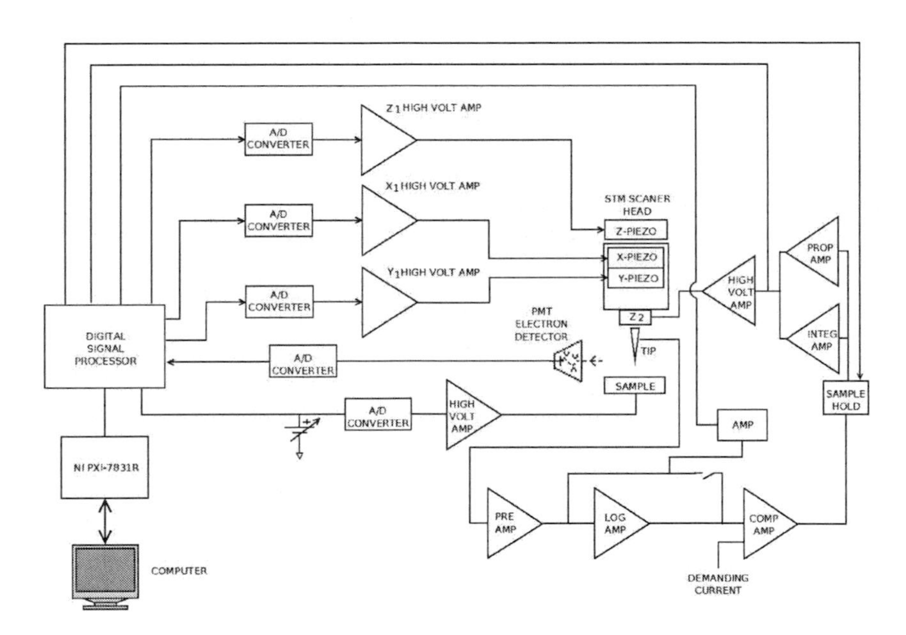

Figure 3.5.: Block diagram of the *NFESEM* electronics.

The requirements detailed in the theory behind the *NFESEM*, demand high electronic sensitivity and stability during the measurements. The electronics must also be compatible with the mechanical parts, *e.g.* piezo, used for the microscope. It is also very important to avoid electrical noise and any ground loops, as well as many other electronic complications. The present set-up consists of a modified, home-built *STM* system that enables high stability and resolution in the measurements, without the aid of sophisticated electronic devices. A block diagram, of the electronics used, is shown in Fig. 3.5.

The ability to apply a high bias voltage is essential to operate the *STM* in *FE* mode, which is limited to a maximum voltage of 10 V to the sample from the computer-controlled digital-to-analog converter (*DAC*). This voltage can be amplified (by a factor of ten or thirty) via a stable high voltage amplifier, which is widely available and allows for other *STM* functions such as voltage modulations for spectroscopy, at a point or the entire scanning region. The *FE* current (voltage) is measured upon passing through a preamplifier (and *I*-*V* converter), after the tip has approached the surface and a current begins to flow between the tip and the sample. Currently a battery-powered National Semiconductor, ultra-low input current amplifier (LMC6001), commonly found in pH probes, analytical medical instruments and electrostatic field detectors, is being used with a gain of 10^7 V/A (or less). Typical measurement *FE* current values range between 1 - 50 nA; therefore, on average, only one electron is between the tip and the sample at a given time [51]. This *FE* current signal is very sensitive and the preamplifier must be placed as close as possible to the tip, so our preamplifier is situated directly at the feedthrough for the *FE* current and is fed into an analog-to-digital converter (*ADC*).

All the signals traveling to and from the computer are processed by a digital signal processor (*DSP*) card, which controls the feedback path of the system in real time. This is half of the feedback loop responsible for measuring of the tunnel (or *FE*) current value that is maintained in *CC* mode, in accordance with section 2.1. Concurrently, the other half keeps the tip height fixed, depending on the set scanning mode, and enables the tip to scan the surface (*i.e.* with the piezos). The relationship between the tip-sample distance and the tunnel current has an exponential dependence, in *CC* mode; hence it is essential to have a logarithmic amplifier. The feedback can be switched between a logarithmic or linear dependence for either the current or voltage, which is controlled via the Nanonis, now SPECS Zurich GmbH, computer program. Subsequently, the current is then checked with the demanded current, as set using the computer program, and the signal from the error between the two is sent to the feedback amplifiers. Finally the signal is directed to the high-voltage amplifier, which controls how the z-piezo reacts. Switching the servo on or off maintains either a constant current between the tip and sample or a constant voltage to the z-piezo respectively. The preferred imaging mode, used for *NFESEM*, is with a constant voltage applied to the z-piezo (*i.e.* *CH* mode).

In addition, the Windows XP operated PC is interfaced with the *DSP* via a NI PXI-8176 (1.26) GHz Pentium III embedded controller and a NI PXI-7831R multifunction RIO with Virtex-II 1M gate FPGA. The board has eight D/A outputs (16-bit resolution, ±10 V, 1 MHz), eight A/D inputs (16-bit resolution, ±10 V, 200 kHz). Three of the output channels are used to control the voltages of the piezos, and one is used for the sample bias voltage. One input channel records the tunneling voltage, one for the *PMT* signal and the rest can be used to read other signals (*e.g.* bias modulations from an internal lock-in-amplifier module). A 16-bit *DAC*/*ADC* is equivalent to 64,000 steps, thus if a voltage of 36 V is applied, it would result in the following positioning resolution:

$$36\text{ V}/64{,}000\text{ steps} = 0.0006\text{ volts/step} = 0.04\text{ Å positioning resolution,}$$

assuming a typical atom is around three angstroms in diameter.

Electrical noise problems are due to ground loops and electromagnetic perturbations produced from power supplies as well as other electronic devices. Preventing electrical noise can be a very tricky. All cables from the chamber and all of the electronic components, used for the measurements, on the rack are all connected to the electronic ground protection of the building via a thick copper cable. The resistance between the connection to the building ground protection and the chamber should be "infinite", when the system ground is not connected to the building ground protection. If there are any ground loops the resistance between the two will be measurable. All connections were also checked for insulation degradation, which can generate noise. Sensitive wires, *e.g.* for *FE* current, are shielded coaxial wires, separated from other wires that can perturb the sensitive signals. Uni-directional wires are also shortened to prevent the formation of coils, which act as an inductor. High voltage wires are separated from low voltage wires, when possible.

3.2. Tip preparation and characterization

"Sharp" field emitters are essential in field ion microscopy (*FIM*), scanning tunneling microscopy (*STM*), and are increasingly being used as electron sources in a number of electron microscopies. The description of the preparation and characterization of the tungsten (*W*)- tips is divided into two parts. First the formation of a "sharp" emitter via electrochemical etching of a cylindrical *W* wire will be explained, which will be followed by a discussion of a high temperature annealing treatment in *UHV*. These processes are more or less standard in *STM*;[3] however the final stage of sharpening is conceptually restricted to *NFESEM*. Furthermore it is possible to gather microscopic information about the emitter using the characterization methods described in section 2.2.10 and 2.2.11.

3.2.1. Ex-situ

The field emitters are fabricated from cylindrical 99.95% polycrystalline *W* wires, with a diameter of either 0.25 or 0.125 mm from Goodfellow Cambridge Ltd. Single crystal *W*-tips of (100)- , (111)- , and (310)-orientation (produced by FEI company) have also been manufactured for *NFESEM* experiments [39]; however the procedure is exactly the same as the 0.125 mm process.

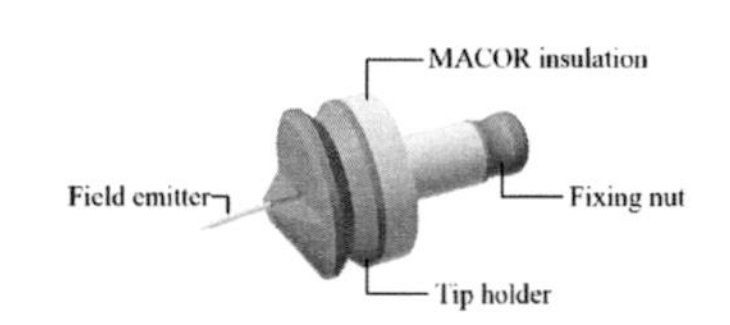

Figure 3.6.: Schematic of field emitter spot-welded to the tip holder.

Initially the surface of the 0.25 mm diameter wire is rough; for this reason it is polished until the end begins to fork. Next the desired length is mechanically cut and spot-welded to the tip holder, such as the one shown in Fig. 3.6. This tip holder is made from titanium, and a MACOR tube is used to reduce the electric field generated at the scanner head.[4] In addition, a titanium fixing nut is attached for electrical contact. Afterwards, the tip holder is mounted on the tip etching stage, shown in Fig. 3.7. The electrolyte used in our electrochemical etching process is a 5 mol/L NaOH solution. Finally the tip is submerged approximately 2 mm into the electrolyte for the pre-etching process. A constant cell voltage of 4.4 - 4.0 Vdc, (for 0.125 - 0.25 mm diameter wire respectively) with respect to the ring-shaped platinum counter-electrode, is applied; thus initiating the etching reaction.

The etching occurs at the air/electrolyte interface; where a concave meniscus of solution is formed around the tip, refer to Fig. 3.8. A necking effect occurs, because the reaction is slower near the air/electrolyte interface. As a result, the current of the circuit, with an automatic switch-off control [53, 54], slowly reduces from (as high as) 11 mA until the "drop-off" of the necked region; consequently "cutting" the circuit, at 1.6 - 1.8 mA.[5] The change in cross-sectional area may also cause the meniscus to shift downwards; therefore a correction of the wire position with the micrometer screw gauge is needed. After the

[3]Please refer to [53] for a review of some of the techniques.
[4]Ideally only the emission region of the tip should be the exposed, conducting surface.
[5]Sometimes even as low as 1.2 mA.

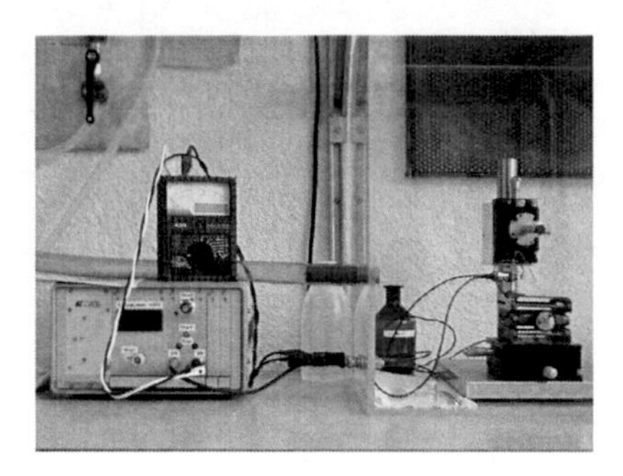

Figure 3.7.: Tip-etching station.

pre-etching, process the tip is further submerged slightly over 1 mm, and the process is repeated. If the current decrease is "smooth" and there are not many current jumps, *i.e.* linear with time, the tip is immediately removed and washed in deionized water at a temperature of approximately 60° C followed by acetone.

It is well-known that faster cut-off (*e.g.* $\sim \mu s$) switches produce sharper tips [54], which is why emitters with a radius of 10 - 50 nm can routinely be fabricated nowadays. Subsequently, the quality of the emitter is verified via an optical microscope. W-tips resulting from a successful electrochemical etching process are quickly transferred to the preparation chamber of the UHV system for a further annealing treatment.

Although the electrochemical etching process is quite simple, it generates unwanted byproducts from the reaction [54]:

$$\begin{array}{rll} \text{Cathode} & 6H_20 + 6e^- & \rightarrow 3H_2(\text{g}) + 6OH^- \\ \text{Anode} & \text{W(s)} + 8\text{OH} & \rightarrow \text{WO}_4^{2-} + 4H_20 + 6e^-. \\ \hline & \text{W(s)} + 2OH^- + 2H_2\text{O} & \rightarrow \text{WO}_4^2 + 3H_2(\text{g}) \end{array} \tag{3.2}$$

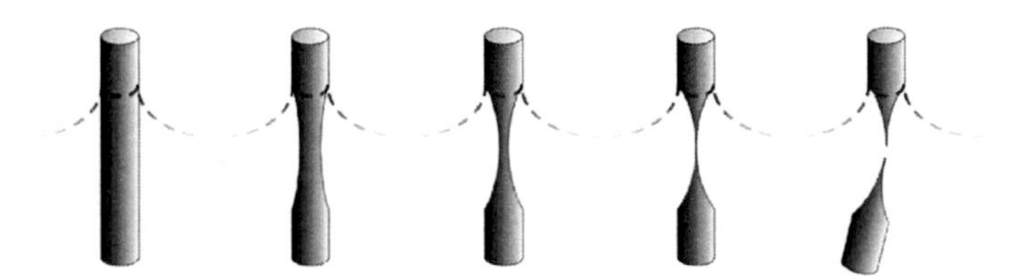

Figure 3.8.: Various stages of the tip etching process.

Furthermore, post-reaction analysis of the electrochemically-etched tips reveal additional surface contaminants such as carbon, from within the wire, oxygen, nitrogen, water-soluble sodium tungstate, as well as other tungsten-oxides. However these contaminants can easily be removed by a simple UHV annealing procedure (discussed in the next section), which is advantageous to tungsten, due to its high melting temperature.

3.2.2. In-situ

Though it is necessary to eliminate additional surface contamination in-situ, the field emitters must be introduced into vacuum via the load-lock, as soon as possible. After achieving adequate vacuum conditions, the tips are transported into the preparation chamber; where the emitters receive a heating treatment using electron bombardment from a 0.125 mm diameter *W* filament. The process entails centering the tip within the first loop of the filament, which is located just below a metallic sphere shown in Fig. 3.9. 500 V is then applied between the emitter and the filament, followed by an increase to the desired current through the filament; hence thermally emitting electrons from the filament to the tip. Initially the electron bombardment current, or emission current, is set at 3.0 - 3.5 mA for 45 minutes in an initial, pre-annealing process to remove weakly bound absorbents and degas the tip holder. Typically the pressure increases to a maximum of $1.5 \cdot 10^{-8}$ mbar, before reducing to $1.0 \cdot 10^{-9}$ mbar at the end of the process. This is followed by a one minute flash using a emission current of 5.5 mA, which results in pressure increase less than $3.0 \cdot 10^{-9}$ mbar. The following reaction of tungsten-trioxide with the bulk tungsten occurs during this final step [55]:

$$2\mathrm{W}O_3 + \mathrm{W} \rightarrow 3\mathrm{W}O_{2\uparrow}; \qquad (3.3)$$

where $\mathrm{W}O_2$ sublimes at $800°C$ [56]. The temperature was estimated to be above $2000°C$, which ensures that all oxides are removed. As the current experimental set-up is not equipped with neither *SEM, FEM, FIM*, nor any method to chemically analyze the emitter, we rely on *FE* analysis as a final method to evaluate the emitter quality. Moreover, as the final application is that the tip acts as an electron emission source, the *FE* characteristics are the most important function.

3.2.2.1. Field emission test for tip sharpness

In general, the *FE* test is used to check the sharpness of the emitter as well as detect for residual contaminants. Although the tip sharpness can be estimated via the $I - V_a$ relationship at the tip preparation stage, a more accurate procedure is conducted in the microscope chamber. This preliminary procedure is performed as follows:

Figure 3.9.: The tip preparation stage. The tip is heated via electron bombardment from the coiled filament, and subsequently the *FE* current is measured between the tip and the grounding sphere.

1. The tip is brought near ($<$ 5 mm) a conducting sphere that is held at ground potential (located just above the filament used for electron bombardment in Fig. 3.9);

2. A negative potential is then applied to the tip inducing a positive bias on the conducting sphere; hence field-emitted

electrons from the tip are collected via the conducting sphere;

3. The tip voltage is increased until the FE current reaches exactly 0.1 nA, adjusting the distance accordingly;

4. Step 3) is repeated for various FE currents at the same distance used in the previous step, generating a I-V curve.

In addition, the presence of surface contaminants and oxides can be observed by the stability of the FE current. Consequently the tip is further sharpened, due to the strong, inhomogeneous electric field that compels the apex towards the grounding sphere. However, this is not a trivial procedure, since it is complicated by the thermal competition between the field strength F and the surface tension γ in the surface migration current J_M [57]:

$$J_M \propto \frac{dz}{dt}\left(\frac{\gamma}{r_{tip}} - \frac{\epsilon_0 F^2}{2}\right); \tag{3.4}$$

where the migration is the result of an external electrostatic field and dz/dt is the rate of change of the emitter length. Here the parenthetical term on the left gives rise to tip blunting if it is greater than the field strength term; however tip sharpening occurs when the parenthetical term on the right dominates. Experimental results show that the rate of change of the length is significantly larger than the rate of emitter radius increase [57]. Spherical electrode field emitters have been modeled in a modified STMsystem; where voltages, ranging from 50 - 1000 V, were used [58]. The effect of Joule heating yielded very low temperature increases in the emitter, $e.g.$10°K at 10mA; however, it has been reported that currents exceeding this threshold will melt the tip [59].

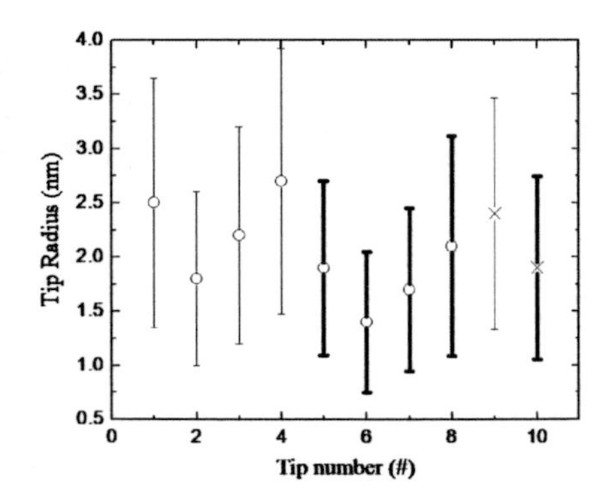

Figure 3.10.: Calculated emission radius for polycrystalline W-tips with work functions of 4.5 eV thin line and 5.25 eV broad line for Fe on Cu (001) (◯) and Au coated HOPG (×) [38].

The emitter is ready to be transferred into the microscope chamber, after the tip cools and the vacuum returns to its base pressure. Finally the emitter is rastered along the sample surface at distances between 100 and 120 nm using biases up to 100 V. This enables additional sharpening of the field emitter via the formation of an asperity smaller than 5 nm, where the electrons are emitted. Thereupon a more accurate $I - V_a$ measurement can be performed at the intended scanning height prior to imaging. The result, of this more accurate measurement, is reported in reference [38] and is summarized in Fig. 3.10.

3.3. Sample preparation and characterization

Sample preparation is an extremely important process in *NFESEM*, as this is a purely surface analysis technique. Only the top layers (< 1 nm) are investigated at primary electron beam energies ranging from 20 - 60 eV (see section 2.3). Furthermore, the sample should be level with no large protrusion, *i.e.* much less than the tip-sample separation. Well-known sputtering and annealing processes are sufficient; however, a fresh surface without contaminants or oxides should be exposed for best results. Surface contamination in the microscope chamber (*i.e.* in a base of $2 \cdot 10^{-11}$ mbar) has significantly suppressed the electron yield after leaving the sample in the chamber for a few days. The following sections details the cleaning process and surface analysis prior to imaging.

3.3.1. Sample cleaning

All samples were prepared immediately preceding imaging; so that the samples were introduced into the microscope chamber less than an hour before a tip approach. Gold-coated **H**ighly **O**rdered/**O**riented **P**yrolitic **G**raphite (*HOPG*) was the first sample investigated, since it is the standard sample used to evaluate the resolution capability in *SEM*. One cycle of sputtering and annealing was applied to the *SEM* resolution test sample: the sample was first bombarded with Ar^+-ions that are accelerated in a 1 kV field, generating a current of $\sim 23\,\mu A$ in a pressure of $\sim 1.0 \cdot 10^{-6}$ mbar for 30 minutes. Subsequently, the sample was annealed via electron bombardment with an emission current, between the W filament and the sample, of 0.5 mA at 250 V for 15 minutes. Initially the pressure increased to $\sim 1.0 \cdot 10^{-9}$ mbar until the heating stage was properly degassed; later the pressure reduced to $\leq 5.0 \cdot 10^{-10}$ mbar.

One cycle of sputtering and annealing was also performed on the sample comprised of a Fe, formed in a shape of a T, deposited on a Cu (001) substrate. Here the same sputtering procedure was used; however, the sample annealing required 90 minutes, with an emission current of 4.0 mA at 500 V. Similarly the sample procedure was applied to the bare Cu (001) substrate.

In regards to the W (110) substrate, a more involved preparation procedure was implemented. First the substrate was annealed at $1000°C$ [6] for 30 minutes. Next oxygen was introduced into the vacuum chamber at a pressure of $4.0 \cdot 10^{-8}$ mbar for 30 minutes, in order to bind with the omnipresent carbon, which migrates to the surface from the bulk. This was followed by an additional annealing in an oxygen free atmosphere for 30 minutes. The substrate was then flashed at a series of temperatures up to $\sim 2000°C$ in order to remove residual CO and CO_2 [60, 61], before finally being flashed at $1700°C$ and cooled.

3.3.2. AES and LEED characterization

Surface-sensitive investigation methods, such as Auger electron spectroscopy (*AES*) and low energy electron diffraction (*LEED*), are essential for the initial characterization

[6] The temperature was determined using an optical pyrometer, Heimann KT19.01, placed outside of the chamber.

of the samples. In particular, *AES* is used to identify the chemical make-up of the surface, which greatly affects the *SE* yield. The most critical are oxides and carbon; therefore the *AES* is meant to detect the concentration of these contaminants. For example, the Auger spectrum of a clean W (110) substrate (using a cylindrical mirror analyzer) is shown in Fig. 3.11a; where the parameters are 2.9 keV primary beam energy, 30 - 700 eV scanning range at a resolution of 0.5 eV, and a 10 ms hold time at each energy interval. The spectrum was averaged over 20 scans of the derivative of the electron count dN/dE, which is plotted as a function of the *SE* energy, to emphasize the peaks. Twenty scans are summed up to give the spectrum, and for better visibility of the peaks corresponding to each element, its derivative is displayed. Although the spectrum is quite noisy, due to instabilities in the electron beam, the elemental peaks are clearly visible. Subsequently *LEED* was performed to determine the crystallinity of the sample, as shown in Fig. 3.11b. Here the unit cell of a clean body-centered cubic structure of the W (110) substrate is visible. The sharpness of the spots and low background intensity indicate that the sample is clean; however, the angle of the camera distorts the symmetry of the spots.

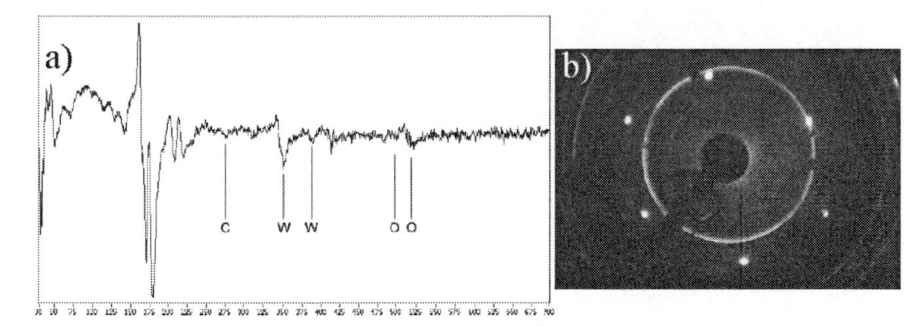

Figure 3.11.: (a) Auger spectrum of the W (110) substrate. Here, the derivative of the electron count dN/dE is plotted as a function of the ejected electron energy (in eVs) to enhance peak visibility. All of the contamination peaks are labeled. (b) *LEED* pattern of the W (110) substrate at 58.3 eV.

4. Results and Discussion

The underlying theory of near field emission scanning electron microscopy (*NFESEM*) has thoroughly been explored, as well as the experimental details associated with the functionality of the microscope. At this juncture, the results of various samples will be presented with an emphasis on the resolution capability of *NFESEM*, as this is the focus of the dissertation. The advances in the development *NFESEM* are closely related to the samples investigated; therefore the data will be discussed in chronological order.

4.1. Fowler-Nordheim curve analysis

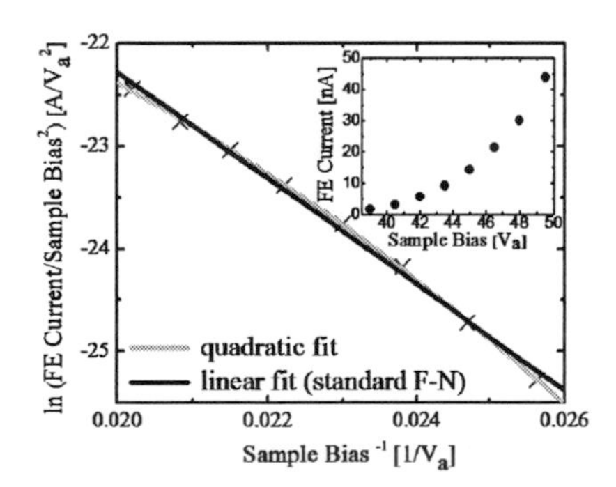

Figure 4.1.: The *F-N* plot × used for the calculations. The linear fit, black line, is a guide for the eyes with a slope of 515.85. The inset shows the original *I-V* data used to determine the *F-N* plot. This *I-V* curve was measured at 150 nm above the sample surface [23].

The field emitter, previously prepared according to the procedure described in Chapter 3, is ready to be transferred into the microscope chamber after the tip cools and the vacuum returns to its base pressure. In a first step, the emitter is scanned along the sample surface (a W (110) substrate, in the present case), in order to form a small asperity for *FE*. The *F-N*-plot of the sharpest field emitter is shown in Fig. 4.1. It is slightly curved; therefore it deviates from the typically linear *F-N*-plots, predicted originally by Fowler and Nordheim. Edgcombe's theory (eq. 2.48 - 2.50), instead, takes into account the sharpening of the emitter described by a "small" (~nm) radius of curvature and produce a slight curvature of the *F-N*-plots. Note that the *F-N*plot is reproducible and the *FE* current is stable, so the measurement error is within the limits of the nonlinear trend.

A similar deviation was observed for *W*-microtips, comprised of one to three atom apex, used in *STM*[62]; however this was attributed to space charge effects. The measurement in Fig. 4.1 was performed at a distance of 150 nm to the substrate in a vacuum chamber with a base pressure of $2 \cdot 10^{-11}$mbar. A quadratic fit, in accordance with eq. 2.48, was performed for the determination of the emission radius. The fit reveals the following results: dS/dV_a = - 39 210 V^2 and S = -410.235 V at 49.5 V. An effective emission radius of less than 1 nm using a *FE* current of 43.8 nA at a voltage of 49.5 V was extracted. This approximation was made using eq. (17) of reference [24] with an average work function of 4.5 eV, given that the *FE* is dominated by the (111)-oriented plane [53]. Bear in mind that the geometrical lateral and vertical resolution are determined more by the tip-surface distance than by the radius of curvature of the emitter. Therefore the slight deviation from linearity of *F-N* plots is certainly a clear sign of a high radius of curvature associated with tip sharpening. This is of interest, because it is used to understand the physical characteristics of the *FE* process, but does not help very much to understand the details of the imaging process. Similar radii of curvature have been obtained in different experiments with different samples.

4.2. Image acquisition

After mechanically placing the tip near the sample surface, an automatic approach can be performed using well-known tunneling parameters for the set current and applied bias, *e.g.* $I_T = 0.15$ nA and $V_a = 200$ mV (for metals). These parameters place the tip within a nanometer of the sample surface, which can be used as a "zero-point" for the vertical positioning of the tip. Upon the retraction of the tip to the desired scan height, the voltage between the tip and the sample may be increased until the desired *FE* current is achieved. If the radius of the emitter is too large, higher voltages will be needed to acquire the same currents as for smaller radii. The smallest (attainable) effective emission radii typically produce the best vertical and lateral spatial resolution. However, tip stability may be problematic in high electric fields and currents, and this is a drawback of using very sharp tips. Clearly, the routine experimental performance will be the result of compromising between various requirements.

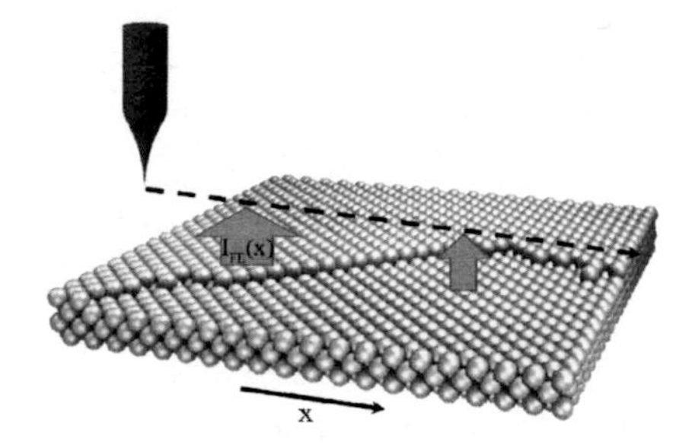

Figure 4.2.: Schematic diagram of the constant height mode.

In order to generate an image, the tip is rastered along the sample surface with the servo switched off, *i.e.* *CH* mode (shown in Fig. 4.2), which is obtained using a sample-and-hold amplifier (please refer to Fig. 3.5). Fig. 4.3 shows the voltage as a function of the tip-sample distance d at a fixed *FE* current $I_T = 0.1$ nA (the starting point for the "green" curve was set with an automatic approach to tunneling distances performed using $I_T = 0.1$ nA and $V_a = 0.1$ V). A fixed *FE* current means that the electric field between the tip and the sample, used to extract the electrons, is kept also constant.

The most remarkable aspect of Fig. 4.3a is the linear increase of the voltage with distance, since it is generally assumed to be constant. This empirical relationship is somewhat surprising, as one would expect the argument of Zuber *et al.* [63] to hold true: $F = V_a/k_f \cdot r_{tip}$ with k_f - a factor accounting for the tip-sample geometry - being $0.5 \ln(4d/r_{tip})$ for $r_{tip} \ll d$ and d/r_{tip} for ($r_{tip} \gg d$). However, *NFESEM* operates in an intermediate range, where $d \approx \mathcal{O}(10 \cdot r_{tip})$. Appendix A.1 provides more of these characteristic curves. There is presently no clear theory explaining this characteristic $V \div d$ curve in the intermediate (transition) regime. One would expect there to be a transition regime, as separation increases, where voltage goes from being linearly dependent on distance d to being nearly constant. To work out what really happens, one would need to do a numerical electrostatic simulation of the whole real geometry of the apparatus, with accurate models of the shape of the emitter and of the support structure on which the emitter is mounted.

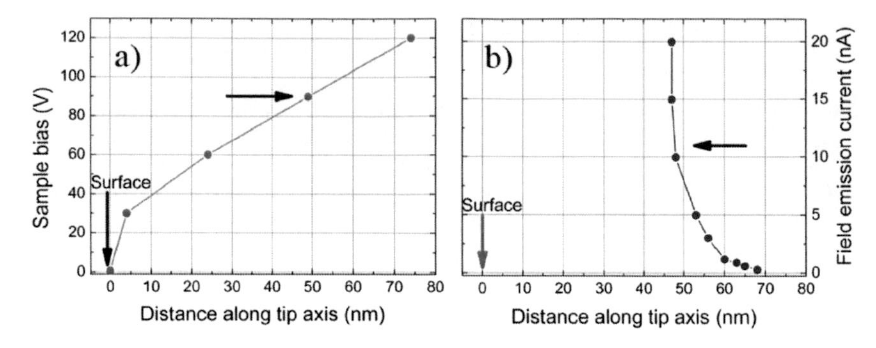

Figure 4.3.: Tip-sample separation dependence. The tip was first brought to the sample surface with $I_T = 0.1$ nA and $V_a = 0.1$ V. (a) Then the bias was increased (green curve) to 120 V, with the current fixed at $I_{FE} = 0.1$ nA. (b) Immediately after the *FE* current was increased (red curve) until the tip ceased to draw nearer to the surface at a fixed voltage of $V_a = 120$ V. The arrows indicate the associated movement of the tip, in accordance with increasing voltage (or current) and the location of the surface.

The *FE* current was then increased (at constant $V_a = 120V$), thus drawing the emitter towards the surface, see Fig. 4.3b ("red" curve). This can be characterized by a strong increase of the *FE* current as the tip-sample distance decreases, which is a direct and straightforward consequence of the "green" curve, if one inserts $F \propto 1/d$ into the Fowler-Nordheim exponent. The strong dependence of the *FE* current on the tip-sample distance is, of course, the justification for the vertical resolution that has been achieved in this experiment. A minute change of the distance at a step edge (about 0.2 nm) varies, in a measurable way, the *FE* current at a typical tip-sample separation of about 25 nm, see the following sections. On the basis of the $F \propto 1/d$-dependence, it is estimated, in fact, from $\Delta(\ln I_{FE}) = \Delta I_{FE}/I_{FE} \approx b\varphi^{3/2}\Delta F/F^2$ and using $b\varphi^{3/2} \approx 65\,\mathrm{V/nm}$, $F \approx 10\,\mathrm{V/nm}$ and $\Delta F/F \approx 0.2/25$:

$$\frac{\Delta I_{FE}}{I_{FE}} = \mathcal{O}(1\%), \tag{4.1}$$

at a step edge, which is the same order of magnitude observed in this instrument (see Appendix A.3).

In order to ensure that the tip is aligned properly, scans are initially made far away from the sample surface, *e.g.* 120 nm, both parallel and perpendicular to the desired scan direction. The slope, in the respective scan direction, is then corrected by applying a voltage to the z-piezo to account for the misalignment, so that images are taken at approximately constant distance between tip and surface (for large tip-sample separation distances). For future developments, it is possible to scan at near video rates in *CH* mode , *i.e.* 20 frames/second [64],[1] with the proper instrumentation. This has not been implemented in the current *NFESEM* system, which scans about one line, in an image, per second. The present sensitivity of the *FE* current to vertical changes allows two

[1]SPECS GmbH (www.specs.de) now produces the Aarhus *STM* with a scan speed in excess of 100,000 pixels/second.

separate images to be taken simultaneously: one using electrons ejected from the surface after interacting with the primary electron beam and one using the *FE* current itself. This produces two topographic maps of the same sample surface [23, 38, 39, 52]. These two images bear a close resemblance, see the following sections, but may exhibit differences depending *e.g.* on the chemical and magnetic contrast.

4.3. Gold coated HOPG

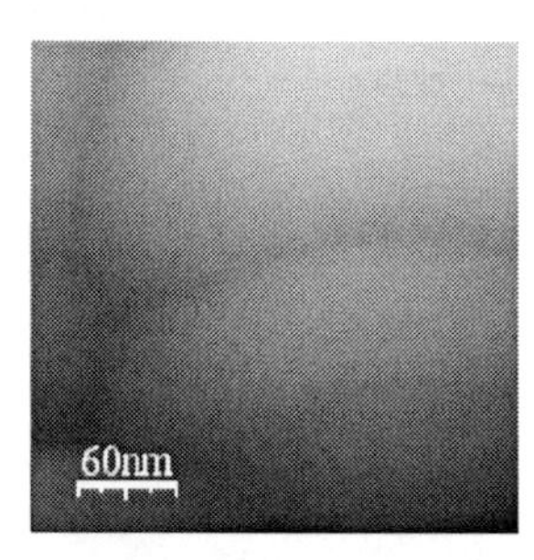

Figure 4.4.: *STM* image of gold grains on HOPG.

The standard procedure to determine the resolution capabilities in *SEM* is, imaging a sample consisting of gold on a carbon substrate, which has been plasma-etched. Typically, hydrocarbon contamination restricts the visibility of grain edges limiting image resolution, due the high acceleration voltages used in conventional *SEM*. Recent advances in sample preparation show that depositing gold on <u>**H**</u>ighly <u>O</u>rdered/<u>O</u>riented <u>P</u>yrolitic <u>G</u>raphite (*HOPG*) and performing a mild annealing prior to imaging, produces uniformly distributed nanometer-scale gold grains and reduces carbon contamination [65]. In accordance with these findings, a similar sample was fabricated to estimate the resolution in *NFESEM*. Here, the sputter-cleaned *HOPG* was first heated to 650°C before depositing an average thickness of 40.8 Å of gold at a slow deposition rate of 0.17 Å/minute. The gold particles were subsequently annealed for 5 minutes to ensure homogeneity. Imaging in normal *STM CC* mode, see Fig. 4.4, reveals tightly-packed spherical gold particles hundreds of nanometers in diameter. All of the *STM* and *NFESEM* gold particle images were performed on this sample, and will be discussed in this section.

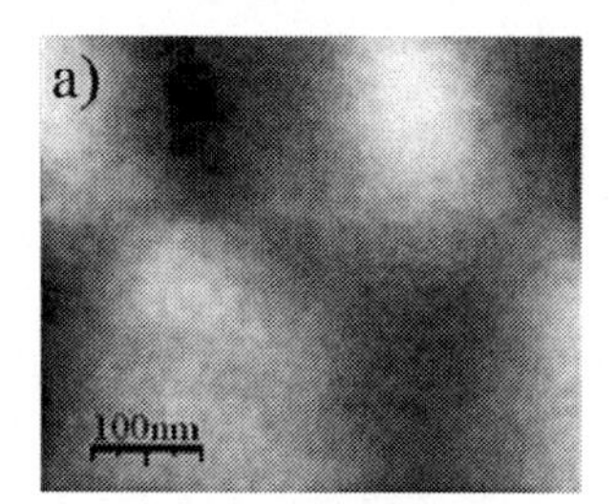

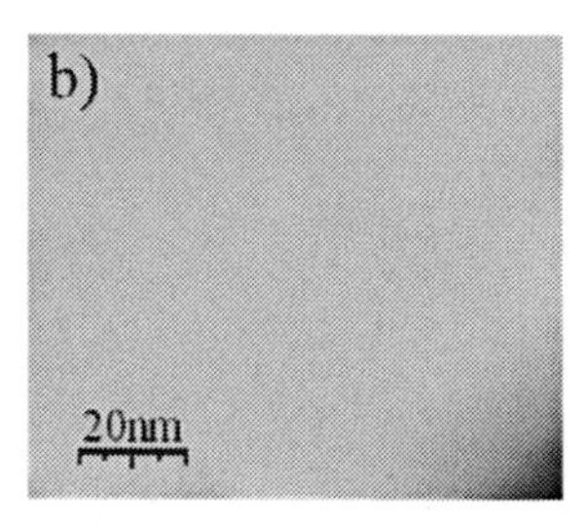

Figure 4.5.: *NFESEM* image of gold grains on *HOPG* recorded at a tip-sample separation of (a) 100 nm and (b) 50 nm.

Upon characterizing the surface of the gold particles, $I - V_a$ measurements were performed at a tip-sample separation of 100 nm in order to determine microscope emitter properties. Subsequently *NFESEM* was performed, using the aforementioned technique in Chapter 3, with the exception of the scintillator type and positioning.[2] Fig. 4.5a was generated by ejected electrons from the sample surface. Here, a primary beam energy of 54.2 eV was employed producing *FE* currents ranging between 30 - 52 nA. Spherically-shaped particles can clearly be seen; however the resolution is quite poor, in comparison to the *STM* image and those attained in high resolution *SEM*. This is however in accordance

[2]This will be discussed later in this section.

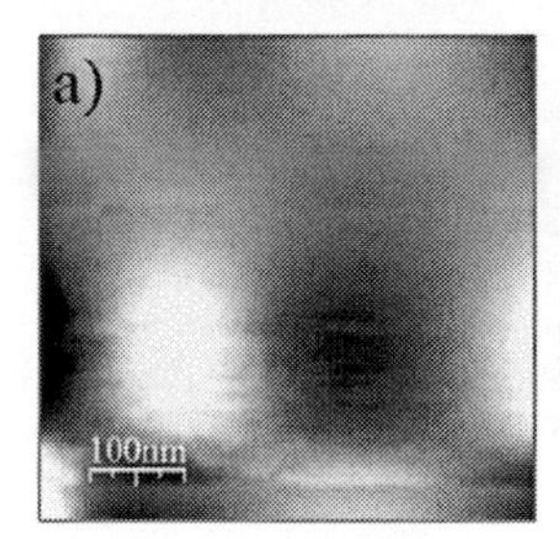

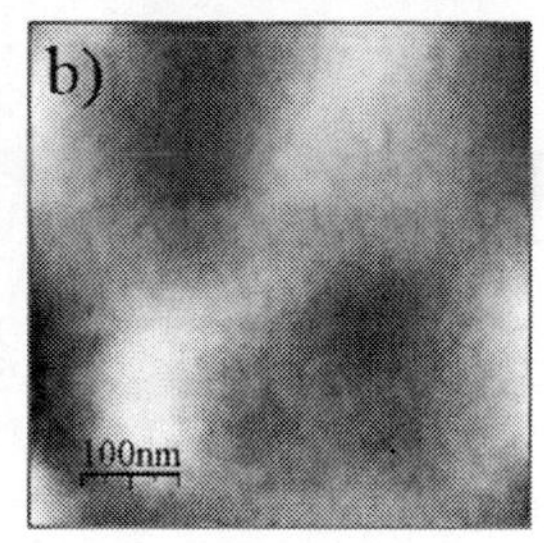

Figure 4.6.: *NFESEM* micrographs of the gold particles using the resultant (a) *FE* current mapping and (b) the electron intensity topography. Here the primary beam energy was 45 eV and the maximum *FE* current was 4 nA, and the measurement was performed at a tip-sample separation of 100 nm using a tip radius of 2 nm (as calculated by eq.2.52).

with the predicted lateral resolution as determined by the formalism used by Sáenz. The emitter radius was estimated using Gomer's version of *F-N* theory, as presented in eq. 2.45. A tip radius of $\sim$ 3.1 nm was deduced, which, in accordance with eq. 2.52, should yield a lateral resolution of $\Delta x \sim 71$ nm. The lateral resolution observed in Fig. 4.5a is perhaps better; however, it is at least the same order of magnitude as the theoretical resolution. In order to be in complete compliance with eq. 2.52, the lateral resolution capabilities should increase accordingly with decreasing tip-sample separation. Therefore an additional image (shown in Fig. 4.5b) was generated at half of the tip-sample separation, *i.e.* 50 nm. The lateral resolution appears to be much better than Fig. 4.5a, as it is clearly demonstrated by the sharp contrast at bottom, right corner of the image. Furthermore, the majority of the image exhibits a very constant white contrast that is caused by the saturation of the detector. This image was produced using a beam energy of 297 eV and an initial *FE* current of 10.1 nA prior to physically crashing into the surface; thus resulting in a blunt field-emitter with no *FE* current at applied biases up to 300 V. A line scan near the gold particle edge reveals a transition region of 20 nm between the completely dark region and the top of the particle, where the detector is saturated.

One must bear in mind that for both of these images and the image in the following section, optimal experimental conditions were not used. The distance between the *SED* and the center of the sample was 9 cm and a standard P-47 scintillator was used. In addition, the detector was mounted at an angle of 22.5°, in reference to the plane of the sample surface; thus requiring high acceleration voltages ranging from 4 - 5 kV to generate the images in Fig. 4.5, 4.6, and 4.7b. In spite of the fact that high resolution images can be produced with this geometry, the scintillator and light guide construction was altered so that the scintillator can be positioned as close as possible to the sample and the surface of the scintillator was centered normal to the sample surface. An actual picture of this set-up is shown in Fig. 3.2a; where the P-47 scintillator was replaced with a $YAlO_3$ perovskite single crystal scintillator (*YAP*), doped with Cerium (Ce^{3+}). The advantages of using the *YAP* scintillator are the following: the quantum efficiency is higher; it produces light at the wavelength for which the photomultiplier tube is most sensitive, 380 nm; and it requires a lower electron acceleration voltage threshold for the production of light, *e.g.* 1.75 kV versus 3 kV using the P-47. These alterations allow for the positioning of the scintillator 2 cm from the sample and reduce the scintillator bias voltage to 2.5 - 3.5 kV.

In Fig. 4.6, the *FE* current (Fig. 4.6a) and the variations in *SED* signal (Fig. 4.6b) are shown and appear to bear a striking resemblance. This shows that the process of producing secondary electrons does not alter the spatial resolution. In summary, these preliminary results indicated that the *SE* yield is strongly dependent on the vertical displacement, since the variations in the *FE* current generate the bulk of the *SE* signal. In fact, this dependence enables the imaging of thin films and single crystal substrates, which will be discussed in the subsequent sections of this dissertation.

4.4. Iron on Cu (001)

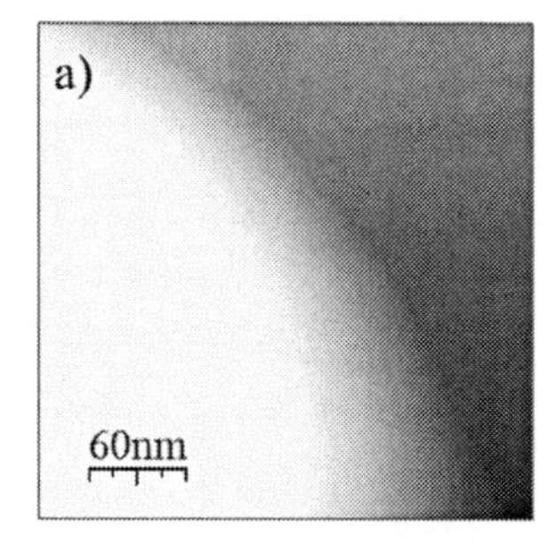

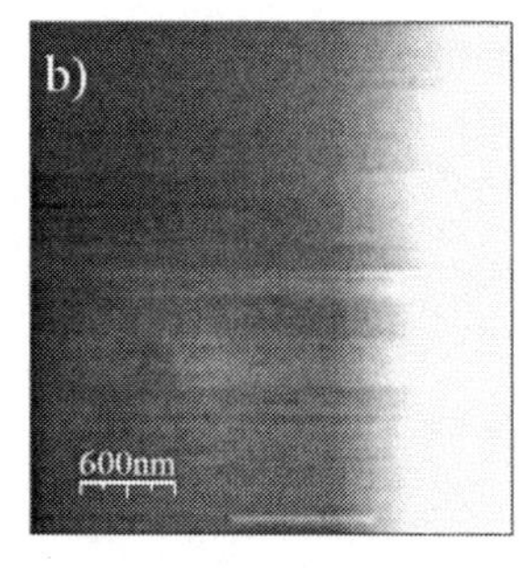

Figure 4.7.: Micrographs of the lateral interface between the edge of an iron shaped T and a Cu (001) substrate imaged via (a) *STM* and (b) *NFESEM*.

The sample consists of an iron deposited on a clean Cu (001) substrate through a T-shaped shadow mask by means of molecular beam epitaxy (*MBE*). Here a Knudsen effusion cell, containing iron wires, was heated via electron bombardment, consequently vaporizing the source. The evaporant flow was then controlled by a manual shutter. Prior to the deposition of the iron evaporant on the Cu (001) substrate, the evaporant was laterally confined by a GaAs shadow mask[3] with a perforated T that was placed directly in front. Structures with sharp edges require that the shadow mask to be placed as close as mechanically possible to the substrate; hence the distance between the two was only a couple millimeters. The substrate was prepared using the method described in section 3.3.1, and the shadow mask was sputtered cleaned prior to iron deposition. An electron bombardment procedure, similar to the sample and the tip annealing, was applied to the Knudsen cell to generate the evaporant using an electron acceleration energy of 1 keV and a current of 75 mA for one hour. The resultant formation is a large iron structure with a sharp lateral interface with the Cu (001) substrate, which is depicted in the *STM* image Fig. 4.7a. A bias of 0.6 V and a set tunnel current of 0.1 nA was used in order to prevent physical tip crashes. The height of the iron structure is approximately 25 nm.

[3] This technique has been applied and described by various members of this group, both past and present, *e.g.* the doctoral dissertation of O. Portmann [66].

Topographic imaging of this lateral interface is also feasible using *NFESEM* on the same sample and the results are shown in Fig. 4.7b. A bias of 1 V and set tunneling current of 0.1 nA was used during the automatic approach towards the iron on Cu (001) surface. These parameters restrict the tip-sample separation; thus preventing tip crashing. After an automatic approach in *CC*-mode the tip was retracted 220 nm to limit tip-sample interactions and perform *NFESEM*. The electrons impinged on the sample with an energy of 180 eV and a *FE* current of 20 nA. Under these parameters, a sharp image contrast is observed; however the intensity of the iron was too great to resolve minuscule surface features. To put it differently, these features may be visible if lower acceleration voltages would have been applied. Moreover, the horizontal lines throughout Fig. 4.7b suggest adsorbate motion between the tip and the sample and/or *FE* current instabilities. Most likely it is a combination of the two. Nevertheless, high resolution imaging of substrates (discussed in the following section) indicate that it is neither the extraction voltage nor the detector positioning, but rather the tip-sample separation, which enables high contrast of minute topographic structures.

4.5. Cu (001)

Granted that the imaging of surfaces with corrugations up to 50 nm is feasible using *NFESEM*, the focus of the present research group is to investigate the fundamental, magnetic properties of ultrathin two-dimensional films.[4] These films are deposited on specially prepared substrates to vary the magnetic anisotropy [66], due to stress at the surface that breaks the symmetry, *e.g.* the easy magnetization axis of 0 -10 monolayers (*ML*)s of iron deposited on Cu (100) rotates from out-of-plane to in-plane [67]. Therefore *NFESEM* must be able to image thin magnetic films on single crystal substrates; hence atomic vertical resolution and nanometer scale lateral resolution is necessary. A topographic *STM* depiction of the Cu (001) can be seen in Fig. 4.8, where the prominent terrace-step edge sequence is a signature of the substrate's crystallinity.

Figure 4.8.: *STM* image of Cu (001) using a bias of 0.2 V and a set tunnel current of 0.2 nA.

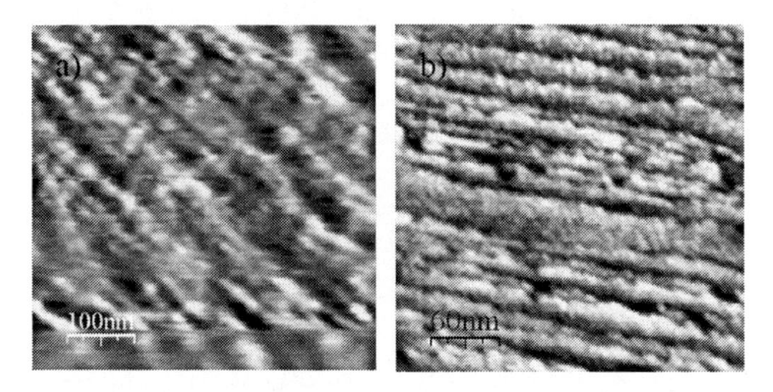

Figure 4.9.: Micrographs of the Cu (001) substrate imaged with (a) *NFESEM* (I_{FE} = 50 nA and E_p= 47 eV) and (b) *STM* (I_T= 0.1 nA and V_B = 0.2 V).

The preliminary data taken with the aforementioned, self-made apparatus, please refer to Fig. 4.9a, are promising. This image was generated by raster scanning the tip at a distance of 75 nm with an estimated emitter radius of 3.7 nm, revealing terraces and single atom steps oriented in a preferred crystalline direction. Line scans of the image show step edges ranging from 12-15 nm wide, which is slightly larger than the *STM*-measured step edges. The surface was cleaned prior to imaging, however the surface still appears to be rough. This difference may be due to the suppression of *SE*s by the strong electric field between the tip and the sample [58] and/or high currents that can damage the sample surface [68]. In fact, the *NFESEM* image is similar to an uncleaned Cu (001), which was imaged using a *STM* (Fig. 4.9b). Another *NFESEM* picture, Fig. 4.10, shows the steps

[4]Refer to www.microstructure.ethz.ch.

and the terraces more clearly as well as the surface deterioration. High beam energies and *FE* currents have been known to induce atomic lattice displacements in copper [69], where beam energies greater than 500 kV and currents ranging between 0.4 - 1 μA in a spot size of 5 μm were used.

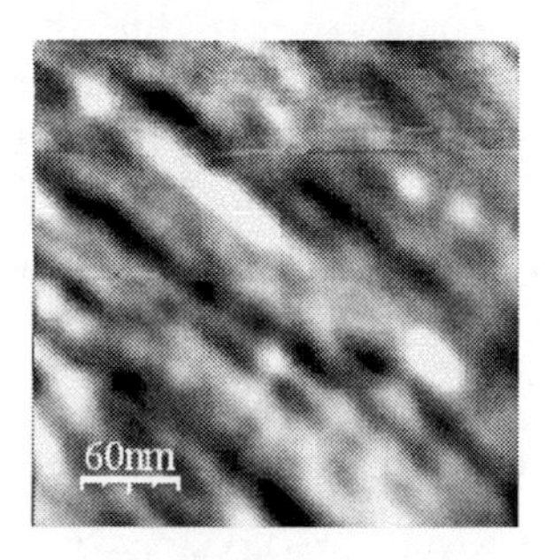

Figure 4.10.: 300 x 300 nm *NFESEM* micrograph of Cu (001) measured at 75 nm (I_{FE} = 50 nA and E_p = 44.5 eV).

More recent studies show that the radiation damage is independent of the beam diameter, and other effects may be the cause [70]. One of the possibilities discussed is electron sputtering, which normally occurs at hundreds of kilovolts, because it requires high current densities *e.g.* $J = 10^5\ {}^{A}/_{cm^2}$. In general, this is only available in transmission electron microscopy (*TEM*), since *SEM* only uses current densities $\approx 10^4\ {}^{A}/_{cm^2}$. These current densities (50 nA at a beam diameter less than 2 nm)[5] are achievable in *NFESEM*, due to the high radius of curvature of the emitter and the sample/electron-extractor geometry. The associated electron energies are far below the sputtering threshold; hence they are unable to displace atoms. Furthermore, the penetration depth of the electrons is small, so that the heat flow is transmitted radially in bulk samples.

J. Cazaux [71] has shown that high current densities, *e.g.* probe size of 1 nm with a *FE* current of 0.4 nA, lead to an electric field of 12 ${}^{V}/_{nm}$ at the edges of the impinging beam. This originates from high energy Auger electrons, sufficient to cause electrical break down [72] and cause lateral migration of ions in insulating material. Electromigration, which describes the mass transport of metal ions in a conductor under the influence of a strong electric field as a result of electron flow [73], has been known to occur on surfaces, if high current densities are used. This is likely the case if copper oxide resides on the surface; however the migration of polycrystalline metals has also been observed in *STM* operating at currents ranging from 1 - 30 nA [74]. It is therefore possible that some copper oxide, residing on the nominally "clean" surface, could cause electromigration [75], since there is no marked difference between gold and copper[6] deposited on *HOPG* [74]. No surface deterioration was observed on the gold particles presented in section 4.3. This will present a challenge when investigating insulating samples, such as biological material.

Currently, the origin of surface deterioration on a Cu (001) substrate under *NFESEM* conditions, is unknown. The detailed investigation of surface damaging processes in *NFESEM* will be a major focus for future research, if one would like to perform *NFESEM* routinely. For the purpose of testing the performance of the instrument, copper substrates were abandoned.

[5]The beam diameter was calculated using the knife-edge beam profile [23, 39, 52].

[6]The surface activation energy for the atomic migration of gold and copper are very similar for the (100) orientation [76].

4.6. W (110)

Figure 4.11.: *STM* image of the W (110) substrate using a bias of 0.2 V and a set tunnel current of 0.2 nA.

NFESEM imaging of the aforementioned samples already indicate that this type of microscopy can potentially be a very powerful tool for surface science investigations on a sub-micrometer scale, yet the most profound results were achieved using a W (110) substrate. These profound results are due to the robustness of *W*, which restricts the adsorbate motion between the tip and the sample. This is due to the fact that *W* has the highest activation energy for surface migration of all the refractory metals [77]. An image of a clean W (110) surface, in accordance with section 3.3.1, has been performed using *STM* (Fig. 4.11) following *NFESEM* imaging (Fig. 4.12). Fig. 4.11 shows rectangular terraces with very sharp and straight step edges, characteristic of the W (110) surface. Moreover, the substrate show no signs of point-like depression defects, and the width of the terraces are more than 100 nm wide. Line scans along the step edges confirm that the steps are of a single atom height, ~ 2Å.

The tip was rastered along the sample surface at 25 nm with the servo switched off, *i.e.*, in constant height (*CH*) mode. The *SED* signal was measured simultaneously with the variations in *FE* current; therefore the measured signals were used to characterize the W (110) substrate. A W (310)-oriented single crystal field emitter was used to image the substrate. The work function of W (310) is 4.25 eV [78], which lowers the tunneling barrier for field-emitted electrons, thus reducing the required field for emission. Accordingly, a lower extraction voltage was applied in comparison to other single/polycrystalline *W*-emitters to achieve a *FE* current of 50 nA [39].

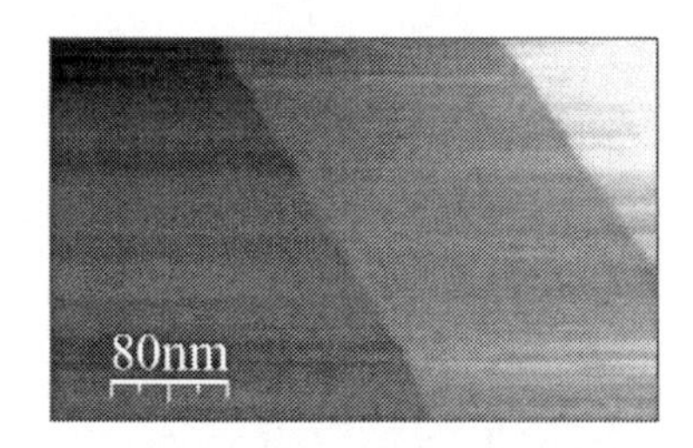

Figure 4.12.: *NFESEM* image of the W (110) substrate measured at 25 nm (I_{FE} = 50 nA and E_p = 28.8 eV).

Even though the image in Fig. 4.12 was not acquired at the same location, it closely resembles the *STM* image in Fig. 4.11; therefore emphasizing the high resolution capability of the microscope. The imaging processes also differ, yet the results are similar, suggesting that *NFESEM* is an alternative method to identify minuscule surface structures. An atomic vertical resolution, which here is based on the small variations of the *FE* current when going from one terrace to the other, cannot be obtained with conventional *SEM*, due to the escape depth of the *SE*, a few nanometers below the surface [26]. Nevertheless, the observation of single atom steps has been reported in *SEM* [79], which requires the use of a special imaging technique for steps ≥ 5Å. In those experiments, either a curved sample is imaged or the sample stage is tilted, in order for the electron beam to strike the sample at grazing incidence. *NFESEM* can easily map the surface of flat, single crystal surfaces with *STM*-like vertical resolution [23, 38, 39, 52], without the need for special sample preparation or positioning of the incident beam.

Note the importance of operating in *CH* mode, which allows detecting the minute change of the *FE* current (typically some tens of pAs on a 50 nA background) when crossing a step edge. Accordingly, the *SE* yield is enhanced by the variations in the *FE* current, and as a result, *FE* current instabilities can be observed in the *SE* image, please refer to the horizontal lines in Fig. 4.12. These *FE* current instabilities may also be due to tip-sample adsorbate motion caused by the electric field between the tip and the sample as well as the high *FE* current. *FE* current instabilities can significantly reduce the quality of the image, as can be observed in Fig. 4.13. Large current spikes in the *FE* current during scanning (Fig. 4.13a) resulted in many discontinuities in the detected *SED* signal (Fig. 4.13b). These sporadic fluctuations are, most likely, the result of improper tip preparation; howbeit this can be remedied by additional tip-flashings at higher temperatures or via field evaporation. This highlights the fact that if the *FE* current were to be pulsed, such as in photo-assisted *FE* [80], the pulse rate would have to be much faster than the rate of *SE* detection.

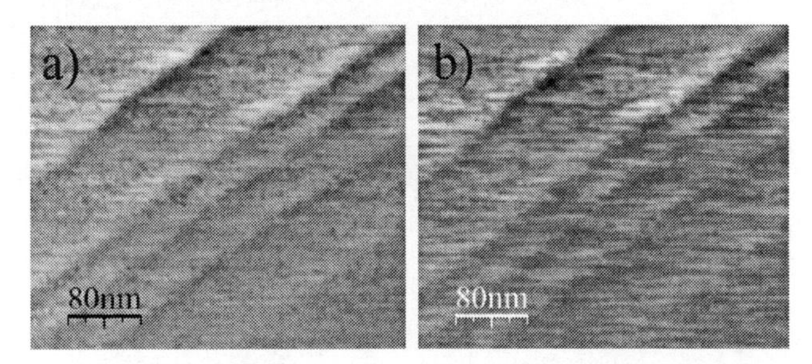

Figure 4.13.: *NFESEM* micrograph of the W (110) substrate measured at 25 nm ($I_{FE} = 50$ nA and $E_p = 34.5$ eV). (a) The *FE* current mapping of the surface was simultaneously recorded with the (b) *SE* topography of the substrate.

4.6.1. Image Enhancement

Although major features of the surface can be observed in the raw image files (see Appendix A.3), the images are enhanced to reveal more contrast and sharper transitions. For example, the correction for the tilt of the sample is often implemented; however this procedure is never performed during the actual imaging of the surface. Large variations of the tip angle, during imaging near the surface, may lead to fracturing of the emitter surface causing the apex to become unstable and eventually separate completely from the emitter. The step edges and terraces of atomically flat surfaces appear as sawtoothlike signals, in the line scan, if sample tilt correction is not adequately implemented, *e.g.* Fig. 4.14. Fortunately, the Nanotec Electronica (WSxM) software [81] allows for the correction of the sample tilt; here the "flattening" parameter enables a line by line subtraction of a

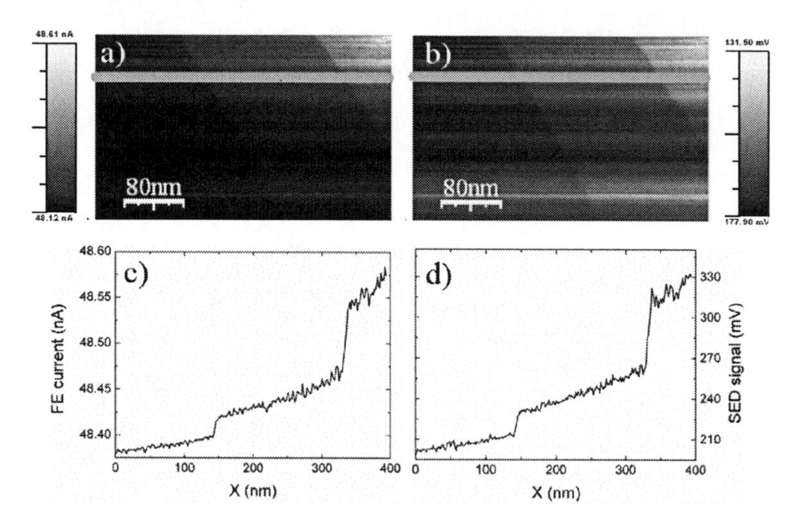

Figure 4.14.: *NFESEM* micrographs of a single crystal W(110) surface. (a) The "raw" signal from the *FE* current mapping and (b) the detected electron signal. The broad grey lines are the regions used for the profiles for the respective images, which are shown for (c) the *FE* current and (d) the *SED* signal (d = 25 nm, I_{FE} = 50 nA and E_p = 28.8 eV).

mean plane or parabola. Subsequently, this results in the image shown in Fig 4.15a. This procedure has been performed on *NFESEM* images, as mentioned in ref. [38], and it is a typical procedure used in scanning probe microscopy.

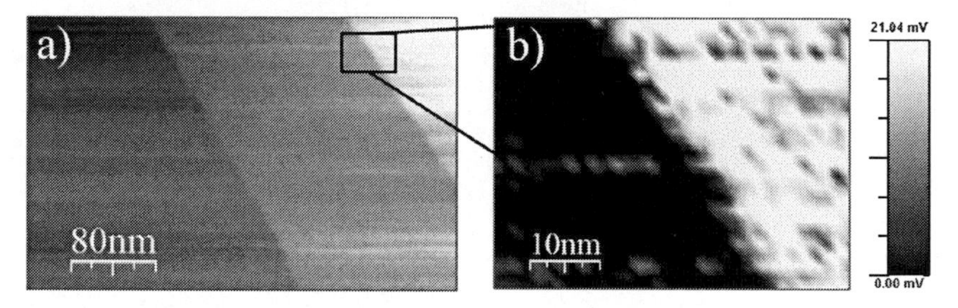

Figure 4.15.: Enhanced NFESEM micrographs of a single crystal W(110) surface. (a) The detected electron signal and (b) the zoom used to determine the lateral resolution at the step edge.

By comparison, it is safe to say that the features observed in Fig. 4.14b are due to the variations in the *FE* current mapping in Fig. 4.14a. In addition, the line profiles of the *FE* current in Fig. 4.14c are almost identical, including the noise, to the detected *SED* signal in Fig. 4.14d. This dependence is a consequence of eq. 3.1, and it emphasizes the collection efficiency of the present *NFESEM* system. Atomic vertical resolution is observed in both images, and the lateral resolution of the *SED* image of the surface can be determined via a line scan along the step edge (see the following section). At the moment, Fig. 4.15a exhibits the highest contrast, and there were no other images of a smaller area made with this level of quality. Hence a zoom of the step edge, Fig. 4.15b, has been used to demonstrate the ultimate *NFESEM* lateral resolution [23, 39, 52]. In order to view the sharp lateral transition at the step edge, the contrast of the image was modified in such a way that the lower terrace was offset to the "0" of the line profile and the upper terrace as the maximum.

4.6.2. Lateral resolution

The surface of a single crystal substrate, *e.g.* W (110), provides an ideal knife-edge to test the resolution, on account of the atomically "sharp" step edges. A series of line profiles were used to determine the lateral resolution for various single crystal and polycrystalline field emitters [39]. Typically the lateral resolution for polycrystalline W-emitters is less than 5 nm [38], but the lateral resolution for the W (310)-oriented tip is about 2 nm [39]. The line scan in Fig. 4.16a, which is the average of several profiles over the step edge (indicated by the black rectangle in Fig. 4.15a, *i.e.* the entire zoom in Fig. 4.15b, was used to determine the lateral resolution. The *FWHM* of the Gaussian fit, to the first derivative of the step edge profiles in Fig. 4.16b, yields a lateral resolution of 1.8 nm.

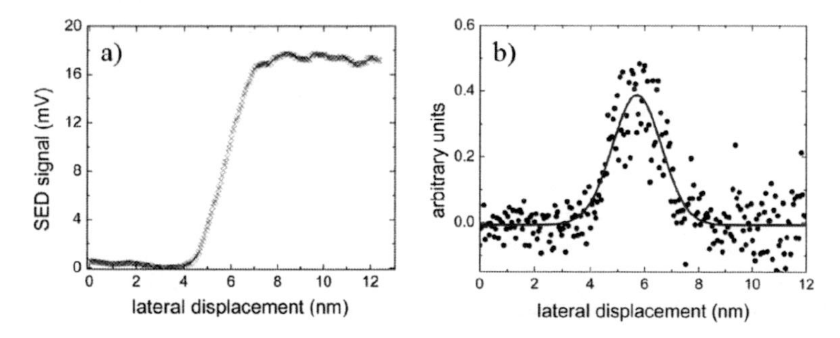

Figure 4.16.: *NFESEM* lateral resolution. (a) Average of a series of line profiles used to calculate (b) the first derivative that is fit to a Gaussian function (red line).

4.7. MBE grown ultrathin Fe films on W (110)

The aim of this section is to show the feasibility of imaging more complex surfaces, for instance those obtained from molecular beam epitaxy (*MBE*) growth of Fe on W (110), which display controlled roughness on a variety of lateral and vertical scales. Ultimately, *NFESEM* will be used for imaging the magnetic contrast of thin films and surfaces.

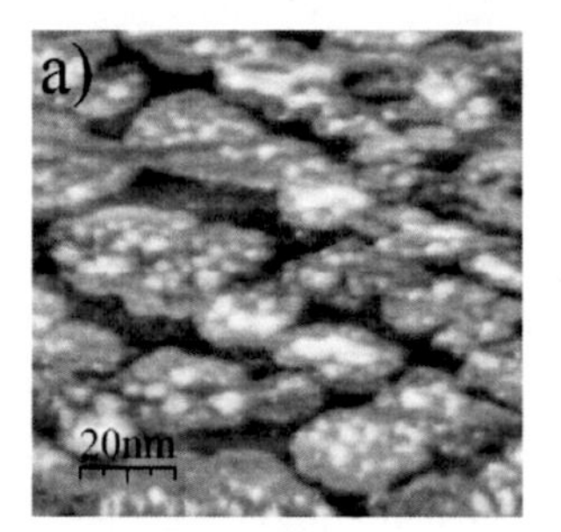

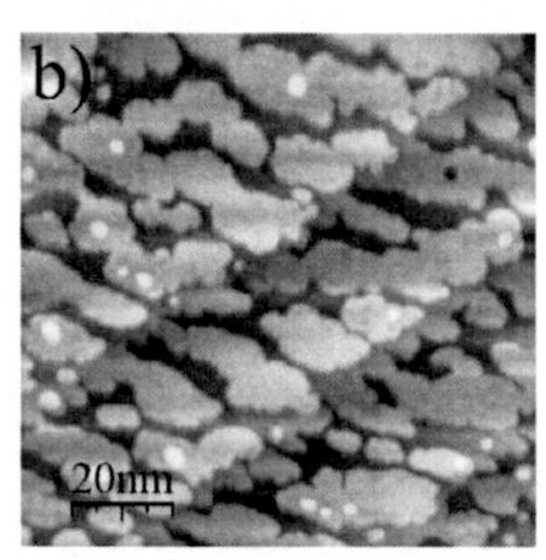

Figure 4.17.: *STM* micrographs of Fe on the W (110) substrate (I_T= 0.15 nA and V_B = 0.2 V). In (a) the dark regions correspond to the *W*-substrate. The first, second, and a large amount of the third Fe layer (brightest spots) are also visible. The coverage corresponds to 1.8 *MLs*, using Auger spectroscopy. In (b) the third layer is less pronounced, indicating that the amount of Fe deposited is less than in (a). The coverage corresponds to 1.1 *MLs*. Two monoatomic steps are seen to cross the image along its diagonal in (b).

Ultrathin films of Fe have been grown on W (110), which has been prepared and characterized following the procedure in section 3.3.1. After the W (110) preparation, ample time was used for substrate cooling (30 - 60 minutes). The iron was grown by *UHV MBE*, see Fig. 4.17. Typically, the layer height is 2.25 Å, as confirmed by line profiles made by *STM*.

After growth, the sample was transported to the scanning Kerr chamber for magnetic measurements. Longitudinal magneto-optical Kerr effect (*MOKE*) detected a square hysteresis loop for a magnetic field applied along the in-plane, $\langle 1\bar{1}0 \rangle$-direction to be magnetized, see *e.g.* Fig. 4.18. No hysteresis was detected for samples with a thickness less than ~2 *MLs*.

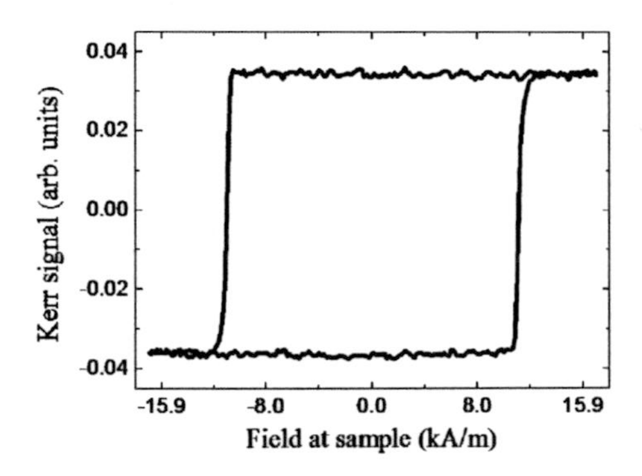

Figure 4.18.: Hysteresis loop of a ~3 *ML* film of Fe on W (110) obtained with a spatially resolved *MOKE* at the center of the sample.

Both *NFESEM* and *STM* were performed on selected Fe coated samples, shown in Fig. 4.19 and 4.20. The samples depicted in Fig. 4.19 and Fig. 4.20 consist of 1.8 and 3.2

MLs of Fe, respectively. In both images, at least four levels of contrast are recognizable, indicating the presence of three Fe layers and a visible W (110) substrate.

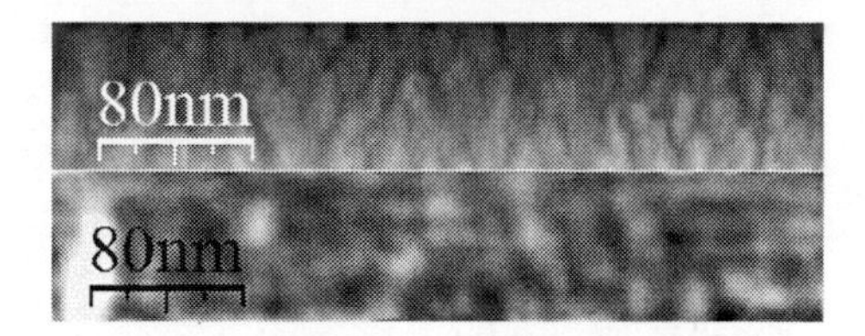

Figure 4.19.: Micrographs of 1.8 *MLs* of Fe on a W (110) substrate imaged with, top image, *STM* (I_T= 0.15 nA and V_B = 0.2 V) and *NFESEM* (I_{FE} = 50 nA and E_p= 50 eV), bottom image.

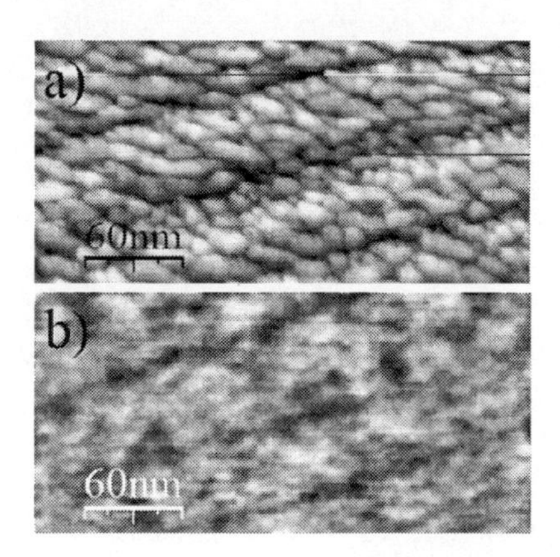

Figure 4.20.: Micrographs of 3.2 *MLs* of Fe on a W (110) substrate imaged with (a) *STM* (I_T= 0.15 nA and V_B = 0.2 V) and (b) *NFESEM* (I_{FE} = 50 nA and E_p= 42 eV).

4.8. Non-ideal images: limitations of the instrument

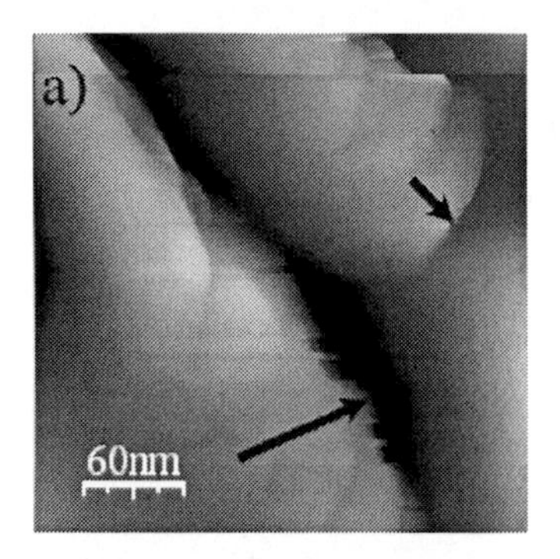

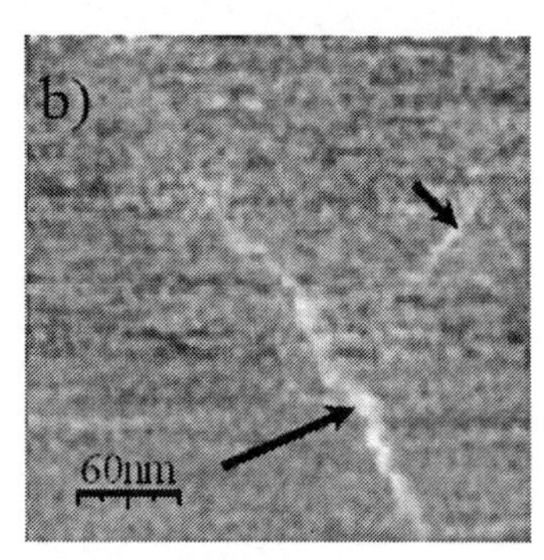

Figure 4.21.: *NFESEM* micrographs of tip debris on the Cu (001). (a) *STM* z-piezo feedback topography (I_{FE} = 0.2 nA and V_B= 20 V) and the (b) simultaneously recorded *SED* signal mapping. The arrows indicate where the tip was retracted during the backward scan.

Figure 4.22.: *NFESEM* image of step bunching on the W (110) substrate measured at 20 nm (I_{FE} = 50 nA and E_p = 44 eV). The inset shows the 2D *FFT*.

NFESEM has proven to be a powerful method to image flat surfaces with nanostructures; assuredly it is not without shortcomings. Here are some more results that should contribute to the understanding of the capabilities and the limits of this instrument. For instance, the *NFESEM* was operated in *CC* mode using a sample bias voltage of 20 V and a constant *FE* current of 0.15 nA, shown in Fig 4.21. This allowed for the concurrent recording of both the z-piezo motion and the *SED* signal. Fig. 4.21a shows the tip debris topography from the z-piezo feedback, and Fig. 4.21b displays the simultaneously measured *SED* signal. Although the depth of the crevice is $\sim 25\ nm$, the *SED* signal only exhibits sharp protrusions caused by the retraction of the z-piezo. One possibility is that the *SE*s could only be detected as the tip was retracted, which was more pronounced during the backward scan. The arrows in Fig. 4.21 show were the large jumps most probably occurred.

NFESEM imaging of single crystal W (110) substrates can easily be performed, due to the large width of the terraces ($\gtrsim 100\ nm$) and atomically sharp step edges. The width of the terrace can be manipulated via the partial pressure of oxygen during annealing

[60]. The short terrace widths in Fig. 4.22, due to insufficient heating after oxidation[7] [60, 84], can reduce the variations in *SED* signal if the electron beam width is of the same order of magnitude of the terrace width. This can be observed in Fig. 4.22, where step bunching can clearly be seen on the W (110) surface with a W (111)-oriented tip. The two-dimensional fast Fourier transform (*FFT*) (Fig. 4.22 inset) reveals a periodic step array with a mean terrace width of 12.5 nm.

Moreover, both the tip and the sample preparation were always assumed to be ideal, but unexpected results often occurred. For example, all of the *W*-emitters received the same heat treatment. In general, this resulted in the desorption of CO, CO_2, and tungsten-oxides. In spite of a reproducible heat treatment, inadequate annealing can lead to faceting of the surface via oxygen surface migration in W (111)-oriented tips [83]. The presence of an oxide on the surface will increase the emitter work function and correspondingly the field required to extract the electrons. This can be observed by the voltage required to perform *NFESEM*, Fig. 4.22 and Fig. 4.23, which is higher than the one used in "ideal" conditions. In addition, the variations in subsequent terrace height is difficult to distinguish in Fig. 4.23. Numerous scans were performed in this same area yielding the same result. A step edge was later confirmed by *STM* measurements at the rift located by the arrow.

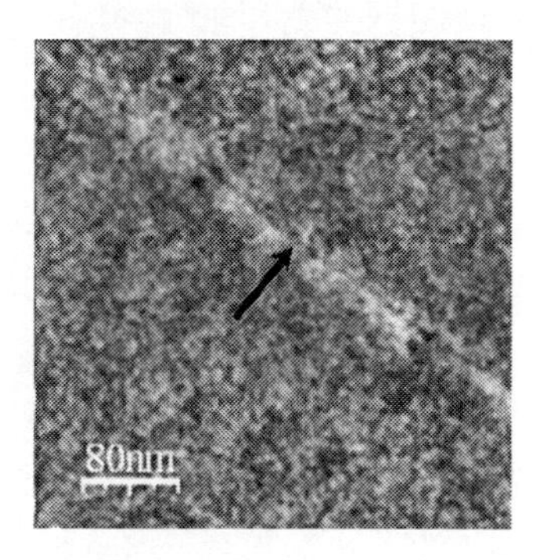

Figure 4.23.: *NFESEM* image of a step edge on the W (110) substrate measured at 20 nm ($I_{FE} = 50$ nA and $E_p = 43.5$ eV). The arrow indicates the step edge.

[7] This was caused by the fusing of the tungsten filament during high temperature flashing.

4.9. Discussion and Outlook

To summarize, a scanning tunneling microscope operating in constant height, field emission mode has been used to generate a well-defined primary electron beam, when positioned in close proximity to a conducting surface. Vertical and lateral spatial resolutions have been determined for "ideal" and "non-ideal" imaging parameters. Deviations from the linearity of the Fowler-Nordheim plot have been detected, which has been attributed to non-planar field emission. C. Edgcombe's field emission theory for curved surfaces was used to determine the emitter radius. A theoretical effort is currently underway to quantitatively understand the details of the imaging process, including the lateral resolution.[8] It could possibly lead to a better understanding of wave-mechanical electron optics for localized electron sources. In addition, studies on the variations of the emitter work function, where higher lateral resolution was observed for single crystal tips with low work function, must be considered. High mechanical stability is also required for the emitters, since they are rastered at small tip-sample separation gaps. Hence the tips remain prone to high electric fields and currents as well as large surface structures. One possibility is replacing the tungsten emitter with a carbon nanotube, because it is mechanically stable and the geometry of the emitter is well-defined. A variety of field emitters will be investigated in the future, which will also be complemented by the optimization of the scanning mechanism.

Although the interaction between the primary electron beam and the sample is similar to *SEM*, the *NFESEM* typically operates at very low primary beam energies (*i.e.* < 60 eV). Electrons are therefore approaching the plasmonic excitation energy in this energy spectrum; where the scattering mechanism becomes dominated by inelastically scattered electron-electron interactions and elastically scattered electron-phonon excitations. This demands a more sophisticated and complete theory to simulate multiple electron trajectories within a given material, in order to determine the *SE* yield. One must also consider the strong electric field between the tip and the sample, to more accurately calculate the number of detectable electrons. The positioning and the functionality of the *SE* detector can be optimized; thus increasing the resolution capabilities of the instrument. Most important, the detector must be able to analyze the spectrum of electrons ejected from the sample. These spectra can possibly lead to additional chemically-based contrast mechanisms; for instance, the up-turn of the inelastic mean free path as a function of energy is material dependent.

The spatial resolution displayed by many of the images exceeds the one predicted by geometrical arguments: $0.7 \cdot [(r+d) \cdot d]^{1/2}$. Various factors may produce this enhancement, for example the emitter shape (see Appendix A.4), the sample topography, detector position, and the chemical make-up of both the tip and the sample. The geometrical limit refers to flat surfaces, but at an infinitely sharp edge, which principally appears in the proximity of monoatomic steps at surfaces. The electric field diverges to infinity, making the step edge a suitable place for the field-emitted electrons to land. Possibly higher resolution capabilities can be achieved by cooling the microscope to cryogenic temperatures; hence increasing the coherence of the emitted electrons and minimizing the adsorbate and surface motion.

There are numerous possibilities for the application of this microscope, due to the

[8] This is in collaboration with Prof. John Xanthakis at the National Technical University of Athens, Greece.

nature of its functionality. *NFESEM* is neither *STM* nor conventional *SEM*; howbeit *UHV* experiments involving nanometer-sized structures on conducting surfaces, limited to either of the aforementioned techniques, can be performed using *NFESEM*. *NFESEM* is also not intended to be coupled with a conventional *SEM*, as in *STM/SEM* systems, but rather to function as a stand-alone instrument. The possibilities for application range from, but are not limited to, electron beam lithography, magnetic recording media, microelectronics, biotechnology, medical studies, magnetic sensors, surface and coating technologies.

One of these applications is currently being implemented, and it will be discussed here.

K. Koike *et al.* [85] have used a "remote source" scanning electron microscopy to image the magnetization distribution at surfaces with high spatial resolution. In *SEMPA*, the remote source produces **unpolarized** electrons: the magnetic contrast results from analyzing the electrons ejected from the surface, in accordance with their spin. In an alternative experiment, spin polarized low electron energy microscopy (SPLEEM), a **spin polarized source** is used, and the scattering asymmetry of elastically reflected electrons is detected [86]. As an extension of these methods to *NFESEM*, a non-magnetic tip could be used to detect the spin asymmetry of some scattered electrons (*NFESEMPA*) or use a magnetic tip and detect the asymmetry of the scattered intensity (*SNFESEM*). Both methods were postulated and considered by D. Pierce [87]. In his account of the feasibility of *SNFESEM*, he confers that there is a variation in scattering intensity on ferromagnetic samples due to a spin-dependent scattering asymmetry. It is estimated that this intensity is greatest for polarized electron beam energies of 100 eV or less, which is within typical *NFESEM* operation parameters. In regards to *NFESEMPA*, it has been shown previously that magnetic signal can be detected using excitations generated by a localized electron source [44, 88].

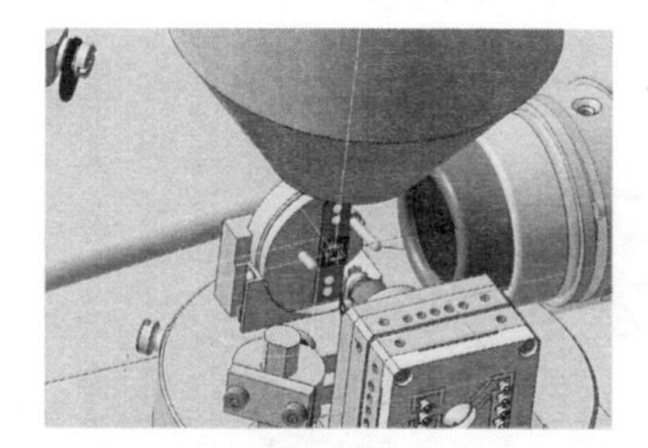

Figure 4.25.: Schematic drawing of the new *NFESEM* system designed to perform both *NFESEMPA* and *SNFESEM*.

The high spatial resolution capabilities for topographic imaging, using the electron intensity of the ejected electron, have already been demonstrated with the *NFESEM*, and the next step is to perform polarization analysis of these electrons. A compact, *UHV*-compatible Mott detector has been fabricated by Ferrovac GmbH modifying the detector detailed in ref. [89]. In this instrument, electrons are extracted from the sample using an extraction voltage ranging between 200 - 600 V. Upon the initial acceleration of the electrons, they are accelerated to 50 kV and are passed through a 90°-deflection, which acts as a low-pass electron energy filter. This is used to select *SE*s with energies 5 eV or less, since they show the highest polarization for the transition metal magnets [87, 90]. These *SE*s are then accelerated towards a gold foil and undergo scattering processes described in references [89, 90]. This detector will be used in combination with *NFESEM* to analyze *SE*, which were excited from an unpolarized electron beam; hence *NFESEMPA*. The measurement principle is shown in Fig. 4.24, and it is similar to standard *SEMPA*, with the exception of the primary electron source. The sample holder is rotatable (see Fig. 4.25), in order to potentially measure the three-components of the magnetization

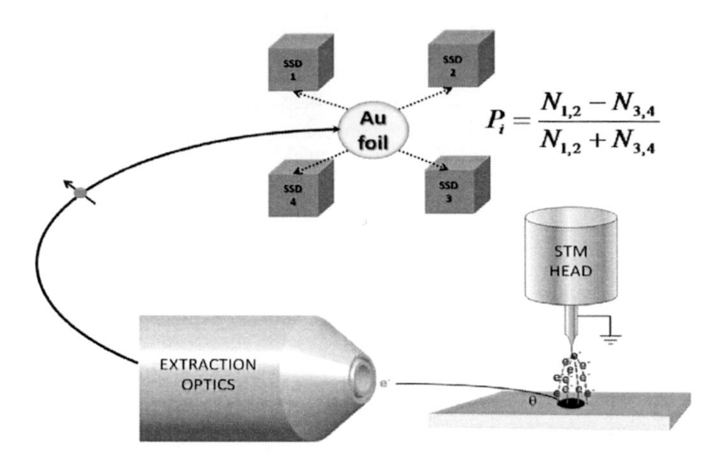

Figure 4.24.: Schematic diagram of the *NFESEMPA*. In this instrument, electrons are excited from the sample surface after undergoing interactions with a primary beam of electrons field-emitted from a sharp tungsten tip. The extraction optics collect all of the subsequent secondary electrons and accelerates the electrons toward a gold foil; where the electrons are asymmetrically scattered spatially, due to spin-orbit coupling with the nuclei of the gold foil. The polarization (P_i) is determined by the normalized difference of the electron count rate on four detectors.

vector, and is more mechanical stable than the previous holder. The Mott detector will be located just above the sample, with the center positioned parallel to the plane of the sample surface, and the central axis of the *SED,* described in section 3.1, will be also be oriented parallel to the surface of the sample (both shown in Fig. 4.25).

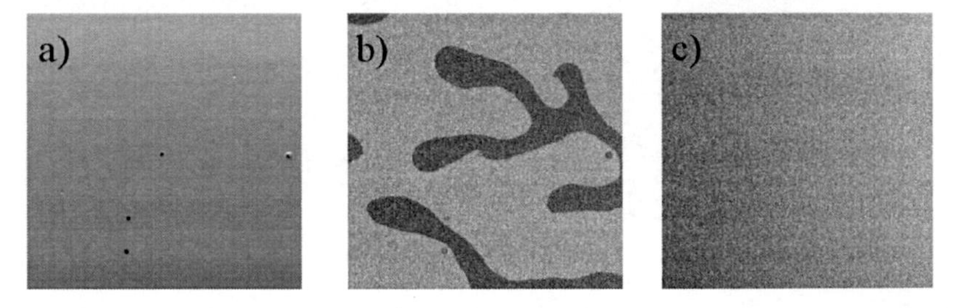

Figure 4.26.: $182\mu m$ x $182\mu m$ micrographs of Fe on Cu (001) using (a) a conventional secondary electron detector, (b) the out-of-plane component of the new Mott detector, and (c) the in-plane component of the new Mott detector.

The Mott detector must be calibrated in order to facilitate an adequate polarimeter efficiency $\in\sim 10^4$ [89], defined as $S_{eff}^2 \cdot I/I_0$; where S_{eff} is the effective Sherman function or asymmetry to be observed at 100% electron polarization. The Mott detector has already been tested using a pre-existing scanning electron microscope, and we were able to image the magnetization vector of a thin iron film deposited on a Cu (001) substrate, please refer to Fig. 4.26. The out-of-plane magnetization vector, Fig. 4.26b, can clearly be separated from the topography of the film shown in Fig. 4.26a and the in-plane component of the magnetization vector, Fig. 4.26c. Moreover, other magnetic samples were measured and a contrast comparable resolution, in comparison to the previous Mott detector, was observed. This Mott detector will soon be integrated with the new *NFESEM* system after it is calibrated.

Bibliography

[1] Yu. V. Sharvin, Sov. Phys. -JETP **21**, 655 (1965).

[2] T. Michlmayr, N. Saratz, A. Vaterlaus, D. Pescia, and U. Ramsperger, *Local-magnetic-field Generation with a Scanning Tunneling Microscopy*, J. Appl. Phys. **99**, 08N502 (2006), T. Michlmayr, N. Saratz, U. Ramsperger, Y. Acremann, T. Bähler, and D. Pescia, *Magnetic Field Generation with Local Current Injection*, J. Phys. D: Appl. Phys. **41**, 055005 (2008).

[3] R.M. Feenstra, *Electronic States of Metal Atoms on the GaAs(110) Surface Studied by Scanning Tunneling Microscopy*, Phys. Rev. Lett. **63**, 1412 (1989).

[4] N. Garcia, *Theory of Scanning Tunneling Microscopy and Spectroscopy: Resolution, Image and Field States, and Thin Oxide Layers*, IBM J. Res. Develop. **30**, 533 (1986).

[5] J. Tersoff and D.R. Hamann, *Theory and Application for the Scanning Tunneling Microscope*, Phys. Rev. Lett. **50**, 1998 (1983).

[6] *Scanning Tunneling Microscopy and Related Methods*, NATO ASI Series E: Applied Sciences Vol. **184**, edited by R. Behm, N. García, and H. Rohrer (Kluwer, Dordrecht, The Netherlands, 1990).

[7] R. H. Fowler and L. Nordheim, *Electron Emission in Intense Electric Fields*, Proc. R. Soc. London, Ser. A **119**, 173 (1928).

[8] R.A. Millikan and Eyring, *Laws Governing the Pulling of Electrons out of Metals by Intense Electrical Fields*, Phys. Rev. **27**, 51 (1926), R.A. Millikan and C.C. Lauritsen, *Relations of Field-Currents to Thermionic-Currents*, Proc. Natl. Acad. Sci. U.S.A. **14**, 45 (1928).

[9] J.R. Oppenheimer, *Three Notes on the Quantum Theory of Aperiodic Effects*, Phys. Rev. **31**, 66 (1928).

[10] L. Nordheim, Z. f. Physik, **46**, 833 (1928).

[11] E.L. Murphy and R.H. Good, *Thermionic emission, field emission, and the transmission region*, Phys. Rev. **102**, 1464 (1956).

[12] R.G. Forbes and J.H.B. Deane, *Reformulation of the Standard Theory of Fowler–Nordheim Tunnelling and Cold Field Electron Emission*, Proc. R. Soc. London, Ser. A **463**, 2907 (2007).

[13] R.G. Forbes, *On the need for a tunneling pre-factor in Fowler-Nordheim tunneling theory*, J. Appl. Phys. **103**, 114911 (2008).

[14] H. Fröman and P.O. Fröman, *JWKB approximation: contributions to the theory*, (North-Holland, Amsterdam, 1965).

[15] L.D. Landau and E.M. Lifschitz, *Quantum mechanics* (Oxford, UK: Pergamon 1958).

[16] W.P. Dyke, J.K. Trolan, W.W. Dolan, and G. Barnes, *The field emitter: Frabrication, Electron Microscopy, and Electric Field Calculations*, J. Appl. Phys. **24**, 570 (1953).

[17] G. Mesa, J.J. Sáenz, and R. García, *Current characteristics in near field emission scanning tunneling microscopes*, J. Vac. Sci. Technol. B **14**, 2403 (1996).

[18] R. Gomer, *Velocity Distribution of Electrons in Field Emission. Resolution in the Projection Microscope*, J. Chem. Phys. **20,** (1952) 1772.

[19] J.He, P.H. Cutler, and N.M. Miskovsky, *Generalization of Fowler-Nordheim field emission theory for nonplanar metal emitters*, Appl. Phys. Lett. **59**, 1644 (1991), P.H. Cutler, J.He, N.M. Miskovsky, T.E. Sullivan, and B. Weiss, *Theory of electron emission in high fields from atomically sharp emitters: Validity of the Fowler-Nordheim equation*, J. Vac. Sci. Technol. B **11**, 387 (1993).

[20] V.T. Binh and J. Marien, *Characterization of microtips for scanning tunneling microscopy*, Surf. Sci. 202, L539 1988.

[21] G. Fursey, *Field Emission in Vacuum Microelectronics*, edited by I. Brodie, P. Shwoebel, and H. Rohrer (Kluwer/Plenum, New York, New York, 2005).

[22] W.P. Dyke and W.W. Dolan, *Field Emission*, Advan. Electron. Electron Phys. **8**, 89 (1956).

[23] T. L. Kirk, O. Scholder, L. G. De Pietro, U. Ramsperger, and D. Pescia, *Evidence of non-planar field emission via secondary electron detection in near field emission scanning electron microscopy*, Appl. Phys. Lett. **94**, 153502 (2009).

[24] C.J. Edgcombe and N. de Jonge, *Deduction of work function of carbon nanotube field emitter by use of curved-surface theory*, J. Phys. D: Appl. Phys. **40**, 4123 (2007).

[25] This should not be confused with *d*, the tip sample separation, see J. Gadzuk and E.W. Plummer, *Field Emission Energy Distribution (FEED)*, Rev. Mod. Phys. **45**, 487 (1973).

[26] L. Reimer, *Scanning Electron Microscopy*, Springer Series in Optical Sciences Vol. **45**, edited by P.W. Hawkes, A.L. Schwlow, T. Tamir, A.E. Siegman, and H.K.V. Lotsch, (second edition Springer-Verlag Heidelberg 1998).

[27] J. Kessler, *Polarized electrons*, Springer Series on atoms and plasmas Vol. **1**, edited by G. Ecker, P. Lambropoulos, and H. Walther, (second edition Springer-Verlag Berlin 1985).

[28] D. Drouin, A.R. Couture, D. Joly, X. Tastet, V. Aimez, and R. Gauvin, *CASINO V2.42 - A Fast and Easy-to-use Modeling Tool for Scanning Electron Microscopy and Microanalysis Users*, Scanning **29**, 92 (2007).

[29] D.E. Newbury, D.C. Joy, P. Echlin, C.E. Fiori, and J.I. Goldstein, *Advanced Scanning Electron Microscopy and X-Ray Microanalysis*, (Plenum Press New York 1986).

[30] E. Casnati, A. Tartari and C Baraldi, *An empirical approach to K-shell ionisation cross section by electrons*, J. Phys. B: At. Mol. Phys.**15**, 155 (1982).

[31] D.C. Joy and S. Luo, *An empirical stopping power relationship for low-energy electrons*, Scanning **11**, 176 (1989).

[32] W.H. Press, B.P. Flannery, S.A. Teukolsky, and W.T. Vetterling, *Numerical Recipes*, 2nd ed. Cambridge University Press (1992).

[33] G. Soum, F. Arnal, J.L. Balladore, B. Jouffrey and P. Verdier, *Monte Carlo calculations on electron multiple scattering in amorphous or polycrystalline targets*, Ultramicroscopy **4**, 451 (1979).

[34] S. Tanuma, C.J. Powell, and D.R. Penn, *Calculations of Electron Inelastic Mean Free Paths 11. Data for 27 Elements over the 50-2000 eV Range*, Surf. Interface Anal. **17**, 911 (1991), C.J. Powell and A. Jablonski, *Evaluation of electron inelastic mean free paths for selected elements and compounds*, Surf. Interface Anal. **29**, 108 (2000).

[35] G. Schönhense and H.C. Siegmann, *Transmission of electrons through ferromagnetic material and applications to detection of electron spin polarization*, Ann. Physik **2**, 465 (1993).

[36] J.J. Sáenz and R. García, *Near field emission scanning tunneling microscopy*, Appl. Phys. Lett. **65**, 3022 (1994).

[37] B. Cho, T. Ichimura, R. Shimizu, and C. Oshima, *Quantitative Evaluation of Spatial Coherence of the Electron Beam from Low Temperature Field Emitters*, Phys. Rev. Lett. **92**, 246103 (1994).

[38] T.L. Kirk, U. Ramsperger, and D. Pescia, *Near field emission scanning electron microscopy*, J. Vac. Sci. Technol. B **27**, 152 (2009).

[39] T.L. Kirk, L.G. De Pietro, D. Pescia and U. Ramsperger, *Electron beam confinement and image contrast enhancement in near field emission scanning electron microscopy*, Ultramicroscopy **109**, 463 (2009).

[40] R. Young, J. Ward, and F. Scire, *The Topografiner: An Instrument for Measuring Surface Microtopography*, Rev. Sci. Instrumen. **43**, 999 (1972).

[41] H.-W. Fink, *Point Source for Ions and Electrons*, Phys. Scr. **38**, 260 (1988).

[42] H.-W. Fink, R. Morin, H. Schmid, W. Stocker, *Low-voltage source for narrow electron/ion beams*, U.S. Patent No. 4954711 (4 Sept. 1990).

[43] H.-W. Fink, *Mono-atomic tips for scanning tunneling microscopy*, IBM J. Res. Develop. **30**, 460 (1986).

[44] P.N. First, J.A. Stroscio, D.T. Pierce, R.A. Dragoset, and R.J. Celotta, *A system for the study of magnetic materials and magnetic imaging with the scanning tunneling microscope*, J. Vac. Sci. Technol. B **9**, 531 (1991).

[45] Ph. Niedermann, N. Sankarraman, R.J. Noer, and O. Fisher, *Field emission from broad-area niobium cathodes: Effects of high-temperature treatment*, J. Appl. Phys. **59**, 892 1986; Ph. Niedermann and O. Fisher, *Application of a Scanning Tunneling Microscope to Field Emission Studeis*, IEEE Trans. Electr. Insul. **24**, 905 1989; V.D. Frolov, A.V. Karabutov, V.I. Konov, S.M. Pimenov, and A. M. Prokhorov, *Scanning*

tunnelling microscopy: application to field electron emission studies, J. Phys. D: Appl. Phys. **32**, 815 1999.

[46] F. Festy, K. Svensson, P. Laitenberger, and R.E. Palmer, *Imaging surfaces with reflected electrons from a field emission scanning tunnelling microscope: image contrast mechanisms*, J. Phys. D: Appl. Phys. **34**, 1849 (2001).

[47] R. Palmer, K. Svensson, P.G. Laitenberger, F. Festy, B. Eves, *Instrument and method for combined surface topography and spectroscopic analysis*, U.S. Patent No. 6855926 15 Feb. 2005.

[48] B.J. Eves, F. Festy, K. Svensson, and R.E. Palmer, *Scanning probe energy loss spectroscopy: Angular resolved measurements on silicon and graphite surfaces*, Appl. Phys. Lett. **77**, 4223 (2000).

[49] J.W. Lyding, S. Skala, J.S. Hubacek, R. Brockenbrough, and G. Gammie, *Variable-temperature scanning tunneling microscope*, Rev. Sci. Instrum. **59**, 1897 (1988).

[50] T.E. Everhart, R.F.M. Thornley, *Wideband detector for micro-micro-ampere low-energy electron currents*, J. Sci. Instr. **37**, 246 (1960).

[51] T.M. Mayer, D.P. Adams, and B.M. Marder, *Field emission characteristics of the scanning tunneling microscope for nanolithography*, J. Vac. Sci. Technol. B **14**, 2438 (1996).

[52] T.L. Kirk, L.G. De Pietro, U. Ramsperger, D. Pescia, *Near Field Emission SEM*, Imaging and Microscopy **11** (1), 35 (2009).

[53] A.J. Melmed, *The art and science and other aspects of making sharp tips*, J. Vac. Sci. Technol. B **9**, 601 (1991).

[54] J.P. Ibe, P.P. Bey Jr., S.L. Brandow, R.A. Brizzolara, N.A. Burnham, D.P. DiLella, K.P. Lee, C.R.K. Marrian, and R.J. Colton, *On the electrochemical etching of tips for scanning tunneling microscopy*, J. Vac. Sci. Technol. A **8**, 3570 (1990).

[55] C.J. Chen, *Introduction to Scanning Tunneling Microscopy*, (New York: Oxford University Press 1993).

[56] I. Ekvall, E. Wahlström, D. Claesson, H. Olin, E. Olsson, *Preparation and characterization of electrochemically etched W tips for STM*, Meas. Sci. Technol. **10**, 11 (1999).

[57] J.P. Barbour, F.M. Charbonnier, W.W. Dolan, W.P. Dyke, E.E. Martin, and J.K. Trolan, *Determination of the Surface Tension and Surface Migration Constants for Tungsten*, Phys. Rev. **117**, 1452 (1960).

[58] M.A. McCord and R.F.W. Pease, *High resolution, low-voltage probes from a field emission source close to the target plane*, J. Vac. Sci. Technol. B **3**(1), 198 (1985).

[59] W.W. Dolan, W.P. Dyke, and J.K. Trolan, *The Field Emission Initiated Vacuum Arc. II. The Resistively Heated Emitter,* Phys. Rev. **91**, 1054 (1953).

[60] M. Bode, S. Krause, L. Berbil-Bautista, S. Heinze and R. Wiesendanger, *On the preparation and electronic properties of clean W(110) surfaces*, Surf. Sci. **601**, 3308 (2007).

[61] U. Ramsperger, *Structural and Magnetic Investigations of Ultrathin Microstructures*, Ph D thesis, Swiss Federal Institute of Technology, Zürich (1996).

[62] V.T. Binh and J. Marien, *Characterization of microtips for scanning tunneling microscopy*, Surf. Sci. **202**, L539 (1988).

[63] J.D. Zuber, K.L. Jensen, and T.E. Sullivan, *An analytical solution for microtip field emission current and effective emssion area*, J. Appl. Phys. **91**, 9379 (2002).

[64] F. Besenbacher, E. Lægsgaard, and I. Stensgaard, *Fast-scanning STM studies*, Materials Today **8**, 26 (2005).

[65] S. Okayama, S. Haraichi, and H. Matsuhata, *Reference sample for the evaluation of SEM image resolution at a high magnification-nanometer-scale Au particles on an HOPG substrate*, J. Electon. Microsc. **54**, 345 (2005).

[66] O. Portmann, *Micromagnetism in the Ultrathin Limit*, Ph D thesis, Swiss Federal Institute of Technology, Zürich (2005).

[67] R. Allenspach and A. Bischof, *Magnetization direction switching in Fe/Cu(100) epitaxial films: temperature and thickness dependence*, Phys. Rev. Lett. **69**, 3385 (1992).

[68] M.A. McCord and R.F.W. Pease, *The effect of reflected and secondary electrons on lithography with the scanning tunneling microscope*, Surf. Sci. **181**, 278 (1987).

[69] M.J. Makin, *Electron displacement damage in copper and aluminium in a high voltage electron microscope*, Phil. Mag. **18**, 637 (1968).

[70] R.F. Egerton, P. Li, and M. Malac, *Radiation damage in the TEM and SEM*, Micron **35**, 399 (2004).

[71] J. Cazaux, *Correlations between ionization radiation damage and charging effects in transmission electron microscopy*, Ultramicroscopy **60**, 411 (1995).

[72] L.W. Hobbs, *Murphy's law and the uncertaintity of electron probes*, Scanning Microsc. Suppl. **4**, 171 (1990).

[73] J.D. Verhoeven, *Electrotransport in Metals*, Metallurgical Reviews **8**, 311 (1963).

[74] M. Ohto, S. Yamaguchi, and K. Tanaka, *Migration of Metals on Graphite in Scanning Tunneling Microscopy*, Jpn. J. Appl. Phys. **34**, L694 (1995).

[75] J. Cazaux, personal communication 27 October 2009.

[76] J.E. Müller and H. Ibach, *Migration of point defects at charged Cu, Ag, and Au (100) surfaces*, Phys. Rev. B **74**, 085408 (2006).

[77] G.A. Cottrell, *Void migration in fusion materials*, J. Nucl. Mater. **302**, 220 (2002).

[78] C.J. Fall, N. Binggeli, Baldereschi, *Theoretical maps of work-function anisotropy*, Phys. Rev. B **65**, 045401 (2001).

[79] Y. Uchida, G. Weinberg, and G. Lehmpfuhl, *Observation of Atomic Steps on Single Crystal Surfaces by a Commercial Scanning Electron Microscope*, Micros. Res. Tech. **20**, 406 (1992) and Y. Homma, *Secondary electron imaging of nucleation and growth of semiconductors for nanostructure fabrication*, Thin Solid Films **332**, 262 (1998).

[80] P. Hommelhoff, C. Kealhofer, and M.A. Kasevich, *Ultrafast Electron Pulses from a Tungsten Tip Triggered by Low-Power Femtosecond Laser Pulses*, Phys. Rev. Lett. **97**, 247402 (2006).

[81] I. Horcas, R. Fernández. J.M. Gómez-Rodriguez, J. Colchero, J. Gómez-Herrero, and A.M. Baro, *WSXM: A software for scanning probe microscopy and a tool for nanotechnology* , Rev. Sci. Instrum. **78**, 013705 (2007).

[82] J.F. Smith and H.N. Southworth, *DETERMINATION OF LAYER THICKNESS BY AUGER ELECTRON SPECTROSCOPY*, Surf. Sci. **122**, L619 (1982).

[83] R. Bryl and A. Szczepkowicz, *The influence of the oxygen exposure on the thermal faceting of W[1 1 1] tip*, Appl. Surf. Sci. **252**, 8526 (2006).

[84] S.M. Zuber, Z. Szczudło, A. Szczepkowicz, Ya.B. Losovyi, and A. Ciszewski, *STM observation of steps and terraces on tungsten (211) surface*, Ultramicroscopy **95**, 165 (2003).

[85] K. Koike and K. Hayakawa, *Scanning Electron Microscope Observation of Magnetic Domains Using Spin-Polarized Secondary Electrons*, Jap. J. Appl. Phy. **23**, L187 (1984).

[86] E. Bauer *et al.*, *Spin-polarized low energy electron microscopy of ferromagnetic thin films* , J. Phys. D: Appl. Phys. **35**, 2327-2331 (2002).

[87] D.T. Pierce, *Spin-Polarized Electron Microscopy*, Phys. Scr. **38**, 291 (1988).

[88] R. Allenspach and A. Bischof, *Spin-polarized secondary electrons from a scanning tunneling microscope in field emission mode*, Appl. Phys. Lett. **54**, 587 (1989).

[89] V.N. Petrov, M. Landolt, M.S. Galaktionov, and B.V. Yushenkov, *A new compact 60 kV Mott polarimeter for spin polarized electron spectroscopy*, Rev. Sci. Instrum. **68**, 4385 (1997).

[90] R. Allenspach, *The attractions of spin-polarized SEM*, Physics World **7**, 44 (1994).

A. Appendices

A.1. Applied voltage versus tip-surface distance

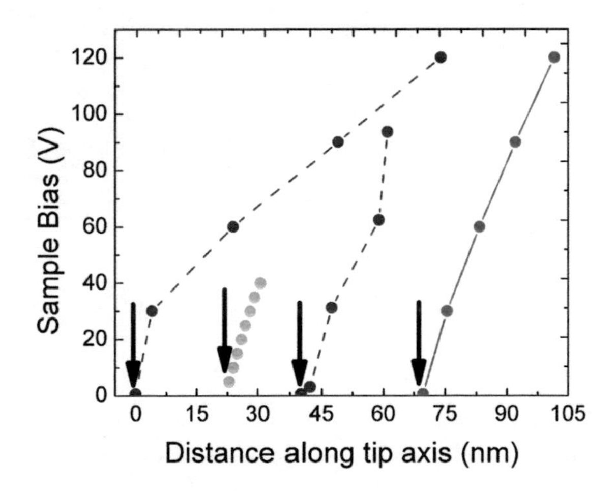

Figure A.1.: Sample bias as a function of the distance along the tip axis for various tips (different colors). The arrows indicate the location of the surface. All curves are obtained with the procedure described in Fig. 4.3a, *i.e.* by keeping the field emission current constant (from left to right: I_{FE}=0.1 nA, 0.2 nA, 0.15 nA, and 0.1 nA). All curves show a strong dependence on the distance between the tip and the surface, which is the basis for achieving subnanometer vertical resolution in images, see section 4.2. Presently, a theoretical model for explaining these characteristic curves is currently underway.

A.2. Dependence of the image contrast on the FE current

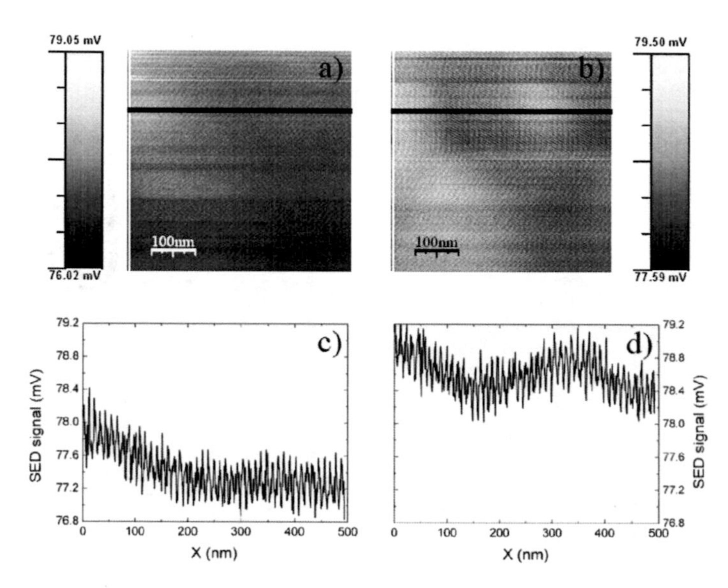

Figure A.2.: Raw *SE-NFESEM* micrographs of gold particles on a *HOPG* substrate. (a) ($d = 100$ nm, $I_{FE} \approx 16$ nA, and $E_p = 60$ eV) and (b) ($d = 100$ nm, $I_{FE} \approx 30$ nA, and $E_p = 64.2$ eV). The line profiles, taken along the black lines of (a) and (b), are given in (c) and (d), respectively. The vertical contrast is clearly improved when the *FE* current is increased from (a) to (b). This is in agreement with the nonlinear current-distance characteristic of Fig. 4.3b ("red" curve), according to which the sensitivity to small vertical changes strongly increases with the *F*E current.

A.3. Resolution results on atomic sized steps

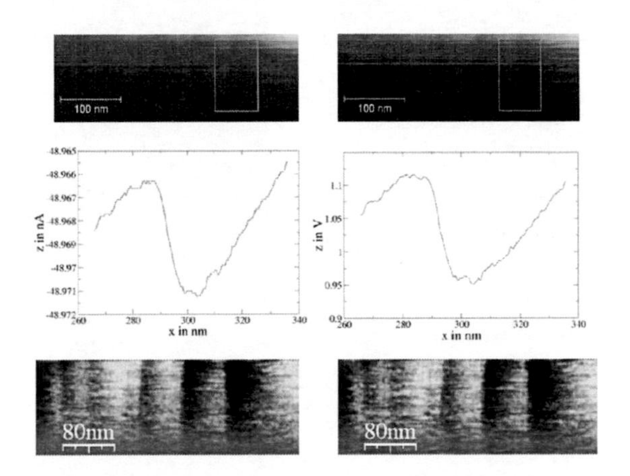

Figure A.3.: Raw *NFESEM* micrographs of a W (110) substrate, (d = 50 nm, $I_{FE} \approx 50$ nA, E_p = 33.2 eV, and scan speed = 425 nm/sec). This surface, based on extended *STM* studies, suggests that the observed contrast is due to mono-atomic steps between flat terraces. The average line scan, taken along the horizontal lines within the marked rectangle, is shown in the middle of the figure. The sawtooth shape of the signal along terraces separated by sharp steps, expected from our *CH* mode, is clearly visible. At the bottom, the raw images are corrected in two ways: first a linear background is subtracted so that the "sawtooth" signal is transformed into a "staircase" signal, better corresponding to a regular, stepped surface. Then, the images are "equalized" – a standard method in image processing which usually increases the global contrast of many images, see U.S. Patent No. 5923383 (13 Jul. 1999). This illustrates the procedure of image enhancement used in section 4.6.1.

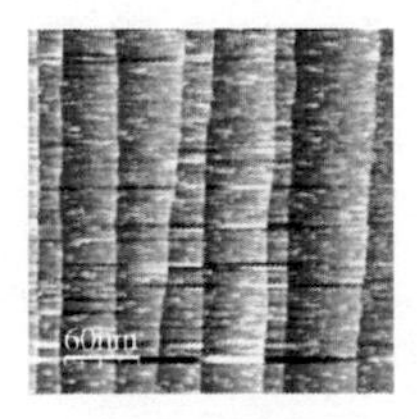

Figure A.4.: The tip-sample separation was decreased for *STM*-imaging in constant current mode after imaging (Fig. A.3) in field emission mode, scanning an area of $300 \times 300 nm^2$ with a tunnel current of 0.2 nA and a bias of 0.2 V. Frequent current jumps occurred during scanning causing the horizontal stripes in the image [38].

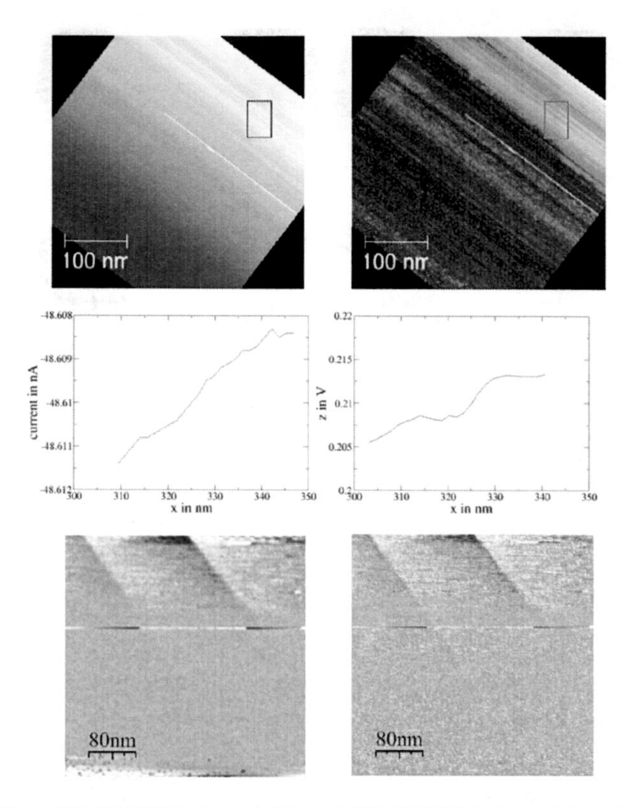

Figure A.5.: Raw *FE-NFESEM* (top, left) and *SE-NFESEM* (top right) of a W (110) substrate, (d = 25 nm, $I_{FE} \approx 50$ nA, E_p = 36.5 eV, and scan speed = 450 nm/sec), rotated to display the steps along the vertical direction, so that the software could perform the averaged line scan within the rectangle (middle of the figure). The bottom two images are enhanced and equalized, see previous Figure and section 4.6.1. The scan was started at the bottom, where the contrast is very faint. Suddenly, the contrast drastically increases, see the thin horizontal stripe (better visible in the bottom images), revealing terraces and step edges, probably because the tip "sharpened". The "sharper" tip appears to be effective for the remaining part of the image, approaching the top.

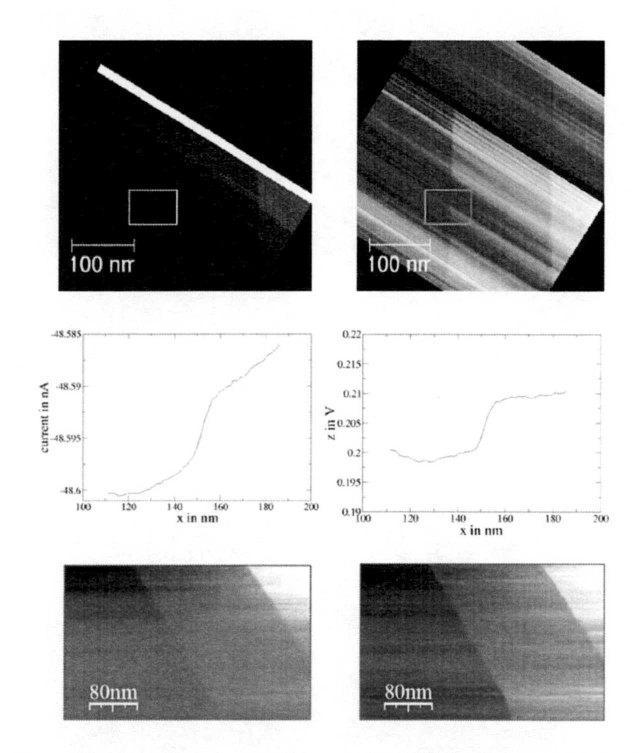

Figure A.6.: Raw *FE-NFESEM* (top, left) and *SE-NFESEM* (top, right) of a W (110) substrate,(d = 25 nm, $I_{FE} \approx 50$ nA, E_p = 28.8 eV, and scan speed = 450 nm/sec). The images are rotated so that the software can compute the average line scans (middle) within the marked rectangle. The bottom two images are enhanced.

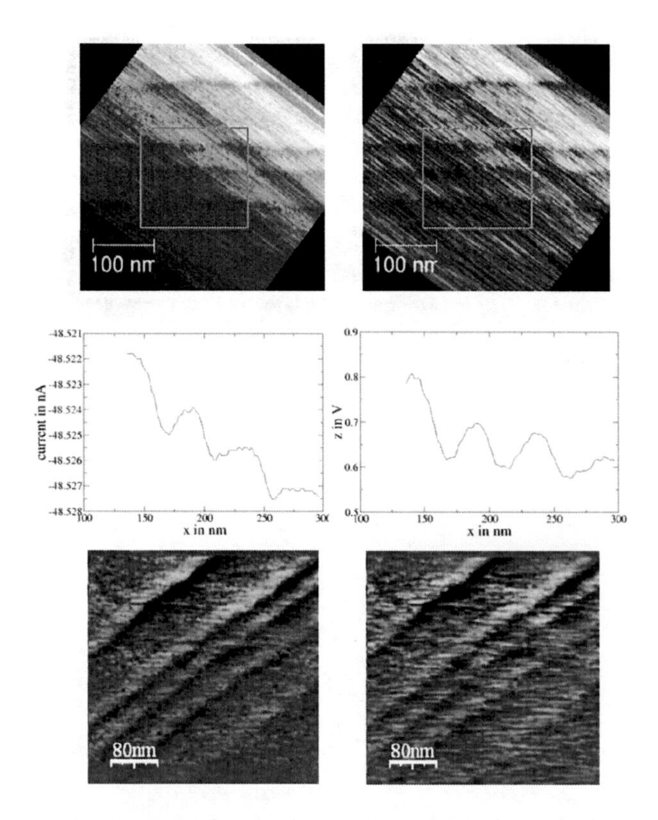

Figure A.7.: Raw *FE-NFESEM* (top, left) and *SE-NFESEM* (top, right) micrographs of a W (110) substrate, (d = 25 nm, $I_{FE} \approx 50$ nA, E_p = 34.5 eV, and scan speed = 450 nm/sec). The images are rotated for the purpose of performing the line average within the marked rectangle. The bottom two images are enhanced and equalized.

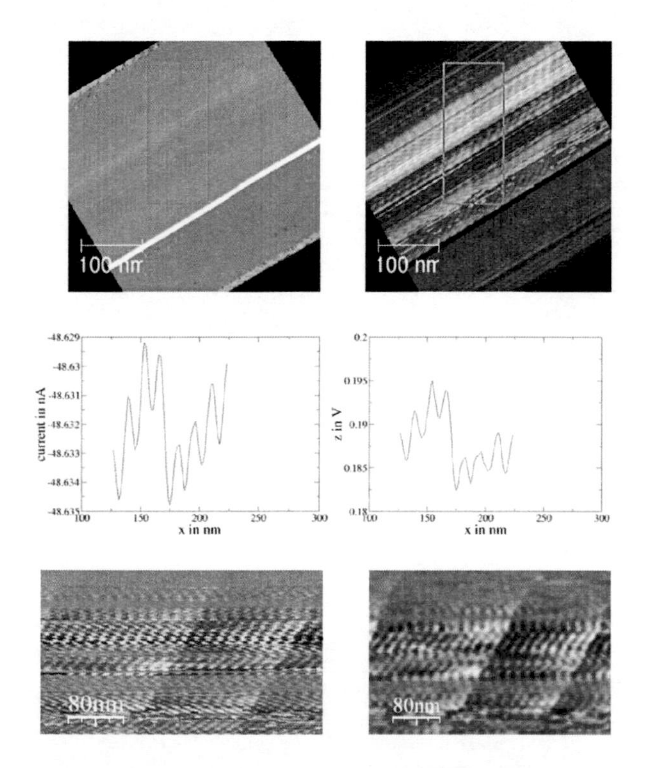

Figure A.8.: Raw *FE-NFESEM* (top, left) and *SE-NFESEM* (top, right) micrographs of a W (110) substrate, ($d = 20$ nm, $I_{FE} \approx 50$ nA, $E_p = 35$ eV, and scan speed = 450 nm/sec). The images are rotated for the purpose of performing the line average within the rectangle, see middle of the figure. The bottom two images are enhanced and equalized.

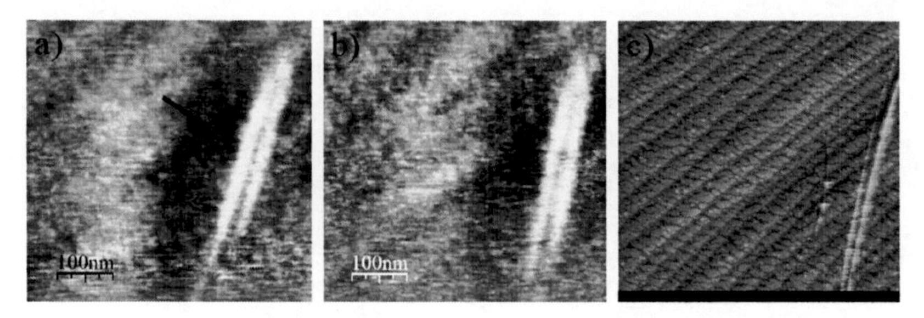

Figure A.9.: Enhanced *NFESEM* micrographs (500 × 500 nm) of 1.1 *MLs* of Fe on a W (110) substrate, (a) ($I_{FE} \approx 20 - 50$ nA and $E_p = 40.7$ eV) upwards scan, (b) ($I_{FE} \approx 20 - 50$ nA and $E_p = 40.7$ eV) downwards scan, and a subsequent *STM* micrograph ($I_T = 0.15$ nA and $V_B = 0.2$ V) performed in the same area. The arrow indicates the iron islands on a terrace. This figure shows the high spatial resolution imaging capabilities of *NFESEM*, as it is compared to a subsequent *STM* image performed in the same scan area. There are two large scratches, most-likely due to the substrate polishing procedure, which help to identify the same locations in all three images. The sample consists of a *MBE* grown Fe on a W (110) substrate, in accordance with section 4.7.

A.4. Application of C. Edgcombe's FE theory for curved surfaces

The radius of curvature is an additional source of error when determining the microscopic properties of the emitter from macroscopic measurables. This has already been addressed in section 4.1, and one proposed solution is to implement C. Edgcombe's theory for non-planar *FE*. Although this theory was designed for a "hemisphere on a post" model; in particular, the theory was developed to describe the emitter properties of a 1 nm thick *CNT*. Here the field enhancement factor γ[1] is expressed as:

$$\gamma \simeq 3 + h/\rho_{tip}; \tag{A.1}$$

where h is the distance from the base of the emitter shank to the apex of the emitter and ρ_{tip} is the emitter radius. Note that this is only applicable when $h \ll d$; where d is the emitter-extractor separation distance.

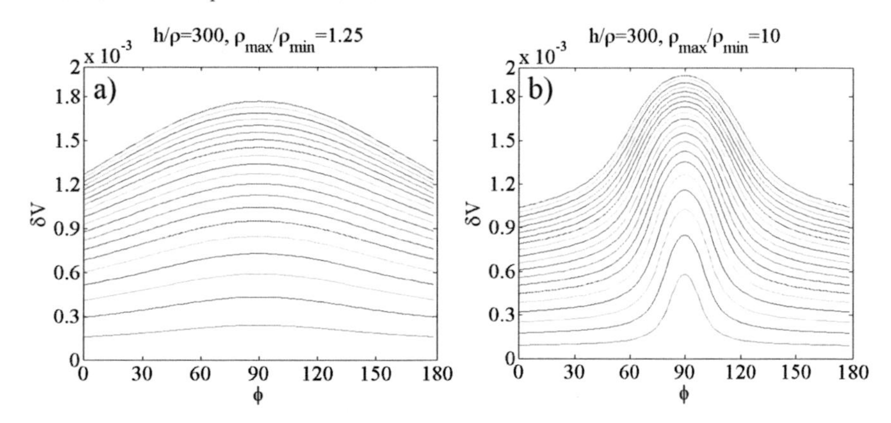

Figure A.10.: Calculated decrease in barrier potential at distance intervals of 10% to the surface of the emitter as a function of polar angle for $h/\rho = 300$. The horizontal axis is numbered in degrees for an emitter apex with an aspect ratio of (a) 1.25 and (b) 10.

In *NFESEM* $h \gg d$; therefore this concept of field enhancement factor is not appropriate. Moreover the emitter is not a "hemisphere on a post," yet nevertheless it was applied to the emitters used in *NFESEM* [23]. Eq. 2.48 was applied in order to determine x, the ratio of the barrier width to the emitter radius. A value of $x \approx 0.64$ was deduced for a "sharp" *NFESEM* emitter; subsequently yielding a value of 3.59 for $\partial f_1/\partial x$. The emitter radius, $r_{tip} \approx 2.4$ Å, was then calculated using eq. 17 of reference [24]. Considering that the radius of curvature of the emitter is of the order of the atomic spacing in tungsten, it is still a reasonable estimate; in accordance with the ultimate lateral resolution of *NFESEM*, ~ 1 nm at a tip-sample separation of 20 nm. In addition, the electron beam divergence angle can be approximated with eq. 14 of reference [24] and the following quadratic fit:

[1]This symbol will be used instead of β, as prescribed by R. Forbes, which has often been associated with the voltage to local field conversion.

$$\Omega = 3.1304A - 1.3909A^2 + 0.3384A^3 - 0.0339A^4, \tag{A.2}$$

where $A \approx 13/SV_a^{-1}$. An angular spread of 1.44 sr was deduced for the present emitter. Not only is this divergence quite large, but also it would be difficult to attain the aforementioned lateral resolution with such a large spread of the beam.

The large solid angle obtained above is, essentially, an artifact of the "hemisphere on a post" model. A more realistic calculation by J.P. Xanthakis, A. Kyritsakis, and G.C. Kokkorakis[2] for an ellipsoidal tip show that the barrier potential falls much faster along the vertical direction ($\phi = 90°$) than in all other directions as the ellipticity of the tip increases, compare Fig. A.10a to Fig. A.10b. Hence as the tip becomes sharper (more non-spherical), the barrier at $\phi = 90°$ becomes weaker, compared to other directions, and the current will be much higher in this direction. This will produce a much less divergent electron beam; or to be more precise, the *FWHM* of the beam will be much less than that of a spherical tip.

[2]This work is currently unpublished, but it will be submitted soon.

CURRICULUM VITAE

Last and first name	Kirk, Taryl
Date of birth	July 14, 1978
Citizen of	The Republic of Trinidad and Tobago and the United States of America

Academic and Research Experience

1992 – 1996	Attended Clovis West High School in Fresno, California
1996	Attained High School Diploma
1996 – 2001	Bachelor of Science studies at the University of California, San Diego with a Major in Physics and a Minor in Philosophy and Mathematics
1998 – 2001	Laboratory assistant in the group of Prof. R. C. Dynes (Department of Physics: Condensed Matter)
2000	Traineeship at I.B.M. Research Labs in San Jose, California
2001	Completed Bachelor of Science degree at the University of California, San Diego
2001	Pre-professional Engineer at I.B.M. Research Labs in San Jose, California
2002	Research and Development Specialist at Gusmer Cellulo Company in Fresno, California
2002 – 2005	Master of Science studies in Physics (Solid State) at the University of Stuttgart, Germany
2002 – 2005	Research Assistant and Master of Science thesis work at the Max Planck Institute for Solid State Physics (Stuttgart, Germany) in the group of Prof. Dr. B. Keimer under the supervision of PD Dr. C. Bernhard
2005	Completed Master of Science degree at the University of Stuttgart, Germany
2005	Research Traineeship at NTT Basic Research Laboratories in Atsugi-city, Japan
2005 – 2006	Researcher at the Swiss Federal Laboratories for Materials Testing and Research (EMPA) in Dübendorf, Switzerland
2005	Guest Scientist at Free University of Berlin, Germany
2007 – 2010	PhD student and teaching assistant at ETH Zurich, Switzerland (PhD thesis at the Laboratory for Solid State Physics in the group of Prof. Dr. Danilo Pescia)

ACKNOWLEDGMENTS

First and foremost, I wish to acknowledge my *Doktorvater*, **Prof. Dr. Danilo Pescia**, without whom this dissertation would not have been possible. He has introduced me to a very interesting project with a number of possibilities, which are not solely limited to physics. I admire **Danilo**'s intuition for scientific writing and his passion for physics. He has provided me with all of the support and, most importantly, the freedom that I needed to achieve significant progressions and direct my own project. He has never impeded my success and he has always stood behind me, even in the direst times. Although I believe this approach should always be the case, it is certainly not the most common practice in science. **Danilo** has given me the opportunity of a lifetime, and I sincerely believe that this may not have been possible anywhere else in the world.

The resources for my doctoral studies were complemented by **Danilo**'s excellent research group, which is comprised of many talents. **Dr. Urs Ramsperger** has an extensive knowledge of all of the laboratory equipment, and has additionally developed a keen sense of problem solving that is greatly appreciated. **Rams** has significantly reduced the amount of time that I would have wasted, trying to develop thorough laboratory procedures. **Dr. Thomas Michlmayr**, who was previously a Ph D student in this group, has also contributed to filling in these gaps and helping me repair equipment. My officemate **Niculin Saratz** has been extremely helpful in many areas of my work, whether it be getting the equipment to work properly or preparing exercises for the classes. He always gives everything 110%, and I wish him the best of luck in his next career move. Sorry that my Caribbean music was so bothersome! Our *Physiklaborant*, **Thomas Bähler**, has not only designed and constructed the secondary electron detector, he is the one responsible for the artistic renditions of the near field emission scanning electron microscope (*NFESEM*), including the cover of this dissertation. I must also mention the *Physiklaboranten-praktikanten*, **Serge Zihlmann**, **Denys Sutter**, and **Michael Nyirö**, who have provided technical support and were also quite helpful to the success of this project.

PD Dr. Rolf Allenspach and **Prof. Phillip First** were the first to observe magnetic contrast using a *NFESEM*-type system with polarization analysis of the secondary electrons. Their work has inspired the development of the *NFESEM*, and I have benefited from fruitful communications with the both of them. The research conducted in **Prof. Richard Palmer**'s group has also helped me to construct the *NFESEM* in the most logical configuration.

Prof. Dr Andreas Vaterlaus was a previous member of our research group and he has founded **Ferrovac GmbH** along with **Urs Maier**. He has recently attained a full professorship at ETH Zürich, and his various transitions in his career have enabled me to weigh my future job options better. **Urs** and his company, **Ferrovac GmbH**, were essential to the development of both the old and new *NFESEM* systems. I wish to especially recognize **Hugo Cabrera**, who is also partially employed by our group and has designed and constructed the new scanning tunneling microscope for a *NFESEM*. *Te doy mis gracias más expresivas.* **Ferrovac GmbH** produces high quality *UHV*-compatible instruments and specializes in both electron polarimetry and scanning probe microscopy. Their system is quite user-friendly and it has made my life easy. Hopefully our continued collaboration will be profitable for them in the future. If it were not for the aforementioned group members, and other affiliates, I would still be at the drawing board.

The controller for the *NFESEM* has been developed by **SPECS Zürich GmbH** (previously **Nanonis**), and we have benefited from our close relationship with them. **Dr. Tobias Vančura** has brought the *NFESEM* to the attention of their parent company **SPECS GmbH**, which has led to a possible collaboration within the scope of a Marie Curie initial training network (*ITN*).

My previous officemates **Dr. Alessandro Vindigni** and **Dr. Oliver Portmann**, though not directly involved in my project, have contributed their professional experiences. Incidentally, it was actually an insightful observation by **Oliver**, which led me to ponder the application of non-planar field emission theory. We have had many interesting conversations, and I have enjoyed their company. **Ale**, **Olivier**, and I had an enjoyable time in Pittsburgh for the APS March meeting 2009.

To all of the previous students that I have had the pleasure of supervising, **Johannes Herrnsdorf**, **Lorenzo De Pietro**, **Johannes Karu**, **Olivier Scholder**, and **Eva Hirtenlechner**, thank you for minimizing the damage to the laboratory! You all have contributed to various stages of my doctoral studies and you have helped me to sharpen my managerial skills. I know all of you will have successful careers and I hope that I was able to contribute to your development in some way. I would especially like to thank **Lorenzo** and **Olivier**, who both did their *Diplomarbeit* and *Masterarbeit* (respectively) with me. You both have provided me with a lot of good data, which was used for my publications and this dissertation. **Lorenzo** has even joined our group as a Ph D student, and I know he will achieve great success developing the magnetic contrast capabilities of the *NFESEM*. I would also like to wish **Anna Stockklauser** and **Mattia Mena** the best of luck for their *Semesterarbeit.*

Für ali vo eu ***müedi Wolkächratzer*** (Tired Skyscrappers): *Mir händ üs ziemli verbessärät in dä letschtä Johrä sogar so viel, dass in däm Johr meh als "nur" dä dritti Platz dinäliegt. Ich hän würklich Spass gha mit Eu z'tschuttä und es hät au zeigt, dass üsäs Team auch usseralb vo dä Physik super funktioniert.*

Çok teşekkür ederim to **Prof. Dr. Mehmet Erbudak,** for all of his help and support. He has taught me a lot about surface science and he has often tested my knowledge, which helped me to learn more. The members of his group, **Dr. Sven Burkardt**, **Dr. Jean-Nicolas Longchamp**, and **Dr. Yves Weisskopf**, have also been helpful with electron diffraction experiments. **Mehmet**, **Sven**, and I also had a memorable time at EASIA09 in Antalya, Turkey.

My work would not be complete, had it not been for a number of individuals in the scientific community who have guided me along the way. Most notably is **Dr. Richard Forbes**, who has pointed out my lack of understanding of the ever-confusing world of field emission theory, during my first presentation about the *NFESEM*. **Richard** has also contributed significantly to the revisions of this dissertation. In addition, he has enlightened me, as well as several others, to the necessary revisions in field emission theory. I am honored to have him as a co-referee, and I look forward to fruitful collaborations in the future. I was also quite fortunate to meet **Prof. John Xanthakis** who has contributed to the theoretical understanding of the lateral resolution of the *NFESEM*. He has taught me a considerable amount about field emission theory, and I consider him to be not only a good collaborator, but also a friend. I believe that our collaboration has produced some fruitful results and I am anticipating a couple of joint publications. **Dr. Chris Edgcombe**'s interest in my work has led to the implementation of his non-planar field emission theory to the emitters for the *NFESEM*. He has many clever ideas about field emission theory and I think that we will both profit from future collaborations, whether it is with or without the *ITN*. My conversations with **Prof. Dr. Jacques Cazaux** have been quite illuminating, and he has given me some interesting ideas for the *NFESEM*. I will certainly remain in close contact with **Jacques** and I hope that I can someday be as helpful to him as much as he has been to me. I have relied heavily on **Prof. Juan Sáenz's** theory for the resolution capability in the "near field" regime, and perhaps the *NFESEM* results will inspire a renewed interest on this topic. My consultations with **Prof. Paul Cutler** and **Prof. Ernst Bauer** have increased my understanding of field emission from small radii and the interactions of a low energy primary electron beam and the sample, respectively.

We are currently collaborating with **Ferrovac GmbH** in a joint project mediated and funded by the Swiss federal innovation promotion agency (*CTI*). It is therefore fitting that I thank **Dr. Vincent Moser** for encouraging me to apply for *CTI* funding, and **Annina Lietha** for reminding me of all my administrative obligations and being helpful (see you on FB!).

The technology transfer at ETH Zürich has played an important role in the patent process, intellectual property rights, and the subsequent commercialization process of the *NFESEM*. The following personnel have all contributed to these various stages: **Dr. Matthias Hölling**, **Dr. Marjan Kraak**, **Dr. Alexa Mundy**, and **Dr. Stefan Lux**. I must also acknowledge the work of the patent attorney, **Dr. Tobias Bremi**. I am also thankful for **Dr. Toshihiko Nagamura**'s interest in the *NFESEM*. *Domo arigatou gozaimashita*! **Toshihiko**-san's ideas are in line with mine, and I believe that my collaboration with **Unisoku co., Ltd.** will generate promising results.

My latest venture was to start a European training program in field emission and Low Energy Electrons for High Resolution Analysis (*LEEHRA*) with **Dr. Christopher Walker**. We have put together an exceptional team of researchers and industrial partners, which consists of **Prof. Mohamed El Gomati**, **Prof. Giovanni Stefani**, **Dr. Ilona Müllerova**, **Prof. Dr. Wolfgang Werner**, **Prof. Crispin Barnes**, **Dr. Stefano Prato**, **Dr. Seyno Sluyterman**, **Dr. Oliver Schaff**, **Dr. Ian Holton**, and the others who have already been mentioned. I still plan on collaborating with many of you, even if our EU proposal is not accepted.

There have been a number of additional individuals who gave me a very strong scientific foundation and have always supported me in the earlier stages of my career in physics. I will only provide a few here: **Charles Chortanian**, **Prof. Robert C. Dynes**, **Prof. Aviad Frydman**, **Dr. Solomon Woods**, **Dr. Andrew Katz**, **Prof. Eric Fullerton**, **Dr. Olav Hellwig**, **Dr. Hans Coufal**, **Prof. Dr. Christian Bernhard**, **Dr. Boris Alexander** (**"Sasha"**), **Dr. Natalia Kovaleva**, **Prof. Dr. Bernhard Keimer**, and **Dr. Hiroyuki Shibata**.

Last but definitely not least, I have to thank my parents for always supporting my decisions, even when my situation in Europe became problematic. It is an honor to be your son, and I am proud that you are my parents. I also thank my brothers, **Nigel** and **Randall**, and the rest of my family members for their love and support. I was quite fortunate to have a family of my own, to provide me with love and support me in all of my endeavors. My family has been both my rock and my inspiration. I know my sons, **Victor** and **Vincent**, will eventually have interesting careers and experiences of their own. Perhaps one of my children will become a physicist. To my lovely assistant, who drew Fig. 2.3, sorry that my studies took up so much time. My wife, **Meike**, is the one responsible for maintaining my sanity and keeping me on the ground.

Although I have been surrounded by numerous talented individuals, and I have received a bountiful amount of love and support from my family, my journey throughout life was made easy by my faith in **God**.

PUBLICATIONS

1) **T.L. Kirk**, "Near field emission scanning electron microscopy," in *Microscopy: Science, Technology, Applications and Education* (Formatex, Badajoz, 2010).

2) **T.L. Kirk**, *STM vs. NFESEM - On epitaxial metal overlayers*, G.I.T. Laboratory Journal Europe **14** (5-6), 18 (2010).

3) **T. Kirk**, U. Ramsperger, D. Pescia, "Low-voltage field emission scanning electron microscope," WO patent 2009100753, Aug. 20, 2009.

4) **T.L. Kirk**, O. Scholder, L.G. De Pietro, U. Ramsperger, and D. Pescia, *Evidence of non-planar field emission via secondary electron detection in near field emission scanning electron microscopy*, Applied Physics Letter **94**, 153502 (2009).

Note: selected for the Virtual Journal of Nanoscale Science & Technology, Nanoscale Science and Technology volume 19, issue 17 (2009).

5) **T.L. Kirk**, L.G. De Pietro, U. Ramsperger, and D. Pescia, *Near Field Emission SEM-Localized Electron Excitation Imaging via SPM*, Imaging and Microscopy **11** (1), 35 (2009).

6) **T.L. Kirk**, L.G. De Pietro, D. Pescia, and U. Ramsperger, *Electron beam confinement and image contrast enhancement in near field emission scanning electron microscopy*, Ultramicroscopy **109**, 463 (2009).

7) **T.L. Kirk**, U. Ramsperger, and D. Pescia, *Near Field Emission Scanning Electron Microscopy*, Journal of Vacuum Science and Technology B **27**, 152 (2009).*

Note: selected for the Virtual Journal of Nanoscale Science & Technology, Nanoscale Science and Technology volume 19, issue 6 (2009).

*Top 20 most downloaded articles February 2009

8) O. Hellwig, **T.L. Kirk**, J.B. Kortright, A. Berger, and E.E. Fullerton, *A new phase diagram for layered antiferromagnetic films*, Nature Materials **2**, 112 (2003).

9) **T.L. Kirk**, O. Hellwig, and E.E. Fullerton, *Coercivity mechanisms in positive exchange-biased Co films and Co/Pt multilayers*, Physical Review B, Vol. **65**, 224426 (2002).

10) A. Frydman, **T.L. Kirk**, and R.C. Dynes, *Superparamagnetism in discontinuous Ni films*, Solid State Communication **114**, 481 (2000).

11) S.I. Woods, A.S. Katz, **T.L. Kirk**, M.C. de Andrade, M.B. Maple, and R.C. Dynes, *Investigation of Nd-Ce-Cu-O Planar Tunnel Junctions Bicrystal Grain Boundary Junctions*, IEEE Transactions on Applied Superconductivity, Vol. **9**, No. 2, Part 3, 3917 (1999).

AWARDS

1) Leipziger Messe and GIT LAB-Award09 for innovative PhD students: **2nd place**

2) International Vacuum Nanoelectronics Conference08: **Best oral presentation**

Bisher erschienene Bände der Reihe
„Applied Electron Microscopy - Angewandte Elektronenmikroskopie“

ISSN 1860-0034

1	Wolfgang Brunner	Beugungsexperimente und Paarverteilungsfunktionen zur Charakterisierung der Struktur von Seltenerdmetall/Eisen-Viellagenschichten im TEM ISBN 3-89722-895-5 40.50 EUR
2	Markus Schneider	Untersuchung des Ummagnetisierungsverhaltens von polykristallinen Mikro- und Nanostrukturen aus Permalloy mit der Lorentz-Transmissionselektronenmikroskopie ISBN 3-8325-0187-8 40.50 EUR
3	Thomas Uhlig	Differentielle Phasenkontrastmikroskopie an magnetischen Ringstrukturen ISBN 3-8325-0752-3 40.50 EUR
4	Martin Heumann	Elektronenholografie an magnetischen Nanostrukturen ISBN 3-8325-0870-8 40.50 EUR
5	Karl Engl	TEM-Analysen an Gruppe III-Nitrid Heterostrukturen ISBN 3-8325-1102-4 40.50 EUR
6	Thomas Haug	Simultaneous Transport Measurements and Highly Resolved Domain Observation of Ferromagnetic Nanostructures ISBN 978-3-8325-1328-3 40.50 EUR
7	Christian Hurm	Towards an unambiguous Electron Magnetic Chiral Dichroism (EMCD) measurement in a Transmission Electron Microscope (TEM) ISBN 978-3-8325-2108-0 40.50 EUR